Water Quality

PROCEEDINGS OF AN INTERNATIONAL FORUM

This is the first volume of a series entitled:

ECOTOXICOLOGY AND ENVIRONMENTAL QUALITY

under the Editorship of Frederick Coulston, Albany, New York, and Friedhelm Korte, Munich

Sponsored by the INTERNATIONAL ACADEMY OF ENVIRONMENTAL SAFETY.

Supported by grant number R803954010 from the UNITED STATES ENVIRONMENTAL PROTECTION AGENCY.

OFFICIAL PUBLICATION OF THE INTERNATIONAL ACADEMY OF ENVIRONMENTAL SAFETY.
Frederick Coulston and Friedhelm Korte, editors.

Water Quality

PROCEEDINGS OF AN INTERNATIONAL FORUM

EDITED BY

Frederick Coulston

Institute of Comparative and Human Toxicology
The Albany Medical College of Union University
Albany, New York

Emil Mrak

University of California
Davis, California

ACADEMIC PRESS NEW YORK SAN FRANCISCO LONDON 1977
A Subsidiary of Harcourt Brace Jovanovich, Publishers

ACADEMIC PRESS RAPID MANUSCRIPT REPRODUCTION

ACADEMIC PRESS, INC.
111 Fifth Avenue, New York, New York 10003

United Kingdom Edition published by
ACADEMIC PRESS, INC. (LONDON) LTD.
24/28 Oval Road, London NW1

Library of Congress Cataloging in Publication Data
Main entry under title:

Water quality.

(Official publication of the International Academy of Environmental Safety)
Forum held in Washington, Sept. 22-24, 1975.
Includes bibliographies and index.
1. Water quality–Congresses. I. Coulston, Frederick. II. Mrak, Emil Marcel, Date III. Series: International Academy of Environmental Safety. Official publication of the International Academy of Environmental Safety.
TD370.W396 363.6'1 76-30602
ISBN 0-12-193150-1

PRINTED IN THE UNITED STATES OF AMERICA

Contents

Speakers and Participants

J. FRANCES ALLEN, US Environmental Protection Agency, Science Advisory Board, Arlington, Virginia 20460

ROBERT ANGELOTTI, Food and Drug Administration, 200 C Street, Washington, D.C. 20036

ALDO H. BAGNULO, Bechtel, Inc., 1620 Eye Street, N.W., Washington, D.C. 20006

ALEXANDRE BERLIN, Health Protection Directorate, Commission of the European Communities, 23 Avenue Monterey, Luxembourg

E. H. BLAIR, Dow Chemical Company, 2020 Dow Center, Midland, Michigan 48640

DOUGLAS E. BUCK, U.S. Borax, 2817 Virginia Avenue, Anaheim, California 92806

JOHN BUCKLEY, US Environmental Protection Agency, 401 M Street, S.W., Washington, D.C. 20460

FRANK A. BUTRICO, Pan American Health Organization, 525 23rd Street, N.W., Washington, D.C. 20037

ROBERT J. BYRON, Calgon Corporation, P.O. Box 1346, Pittsburgh, Pennsylvania 15230

C. O. CHICHESTER, The Nutrition Foundation, 489 Fifth Avenue, New York, New York 10017

LOUIS C. CLAPPER, National Wildlife Federation, 1412 16th Street, N.W., Washington, D.C. 20036

CARL C. CLARK, Citizens' Drinking Water Coalition, 1875 Connecticut Avenue—Suite 1013, Washington, D.C. 20009

H. COSTA, Administrative Manager—Gesellschaft für Strahlen- und Umweltforschung, mbH, 8042 Neuherberg, Ingolstadter Landstr. 1, West Germany

FREDERICK COULSTON, Institute of Comparative and Human Toxicology, Albany Medical College, 47 New Scotland Avenue, Albany, New York 12208

DAVID H. CRITCHFIELD, US Environmental Protection Agency, 401 M Street, S.W., Washington, D.C. 20460

ROBERT F. CURRAN, Ciba Geigy Corporation, Saw Mill River Road, Ardsley, New York 10502

PIERRE Y. DIVET, Calgon Corporation (Chemviron), P.O. Box 1346, Pittsburgh, Pennsylvania 15230

DAVID DONALDSON, Pan American Health Organization, 525 23rd Street, N.W., Washington, D.C. 20037

HAROLD EGAN, Department of Industry, Laboratory of the Government Chemist, Cornwall House, Stamford Street, London, England

D. E. W. ERNST, Institut für Biophysik, Techische Universität, Hannover, Herrenhauser Str. 2, 3 Hannover, West Germany

LOU C. GILDE, Campbell Soup Company, Campbell Place, Camden, New Jersey 08101

LEON GOLBERG, Institute of Comparative and Human Toxicology, Albany Medical College, 47 New Scotland Avenue, Albany, New York 12208

MATTHEW GOULD, Georgia-Pacific, 900 S.W. 5th Avenue, Portland, Oregon 97204

FITZHUGH GREEN, US Environmental Protection Agency, 401 M Street, S.W., Washington, D.C. 20460

H. GREIM, Department of Toxicology, Gesellschaft für Strahlen- und Umweltforschung, mbH, 8042 Neuherberg, Ingolstadter Landstr. 1, West Germany

HANS GYSIN, Ciba Geigy AG, Basel, Switzerland

JOHN D. HALLETT, Shell Oil Company, Houston, Texas 77068

BRUCE HATHAWAY, Environment Reporter, 1231 25th Street, Washington, D.C. 20037

L. HOPKINS, Chevron Chemical Company, 200 Bush Street, San Francisco, California 94104

RILEY D. HOUSEWRIGHT, National Academy of Sciences, 2101 Constitution Avenue, Washington, D.C. 20037

J. T. HUTTON, Vice President and Director of Research, Foremost Foods Company, One Post Street, San Francisco, California 94104

JOSEPH C. JACKSON, A/C Pipe Producers Association, 1875 Connecticut Avenue, N.W., Washington, D.C. 20009

CAROL JOLLY, League of Women Voters, 1730 M Street, N.W., Washington, D.C. 20036

FRED H. JONES, American Bottled Water Association, 1411 W. Olympic Boulevard, Los Angeles, California 90015

JULIAN JOSEPHSON, Environmental Science and Technology, American Chemical Society, 1155 16th Street, N.W., Washington, D.C. 20036

GREGOR A. JUNK, US Energy Research and Development Administration (ERDA), Iowa State University, Ames, Iowa 50011

IWAO KAWASHIRO, Director—National Institute of Hygienic Sciences, Kamiyoga, Setagaya-ku, Tokyo, Japan

THOMAS L. KIMBALL, Executive Vice President, National Wildlife Federation, 1412 16th Street, N.W., Washington, D.C. 20036

VICTOR J. KIMM, US Environmental Protection Agency, 401 M Street, S.W., Washington, D.C. 20460

DONALD G. KIRK, H.J. Heinz Company, P.O. Box 57, Pittsburgh, Pennsylvania 15230

WERNER KLEIN, Institut für Okologische Chemie, Gesellschaft für Strahlen- und Umweltforschung mbH, 5205 St. Augustin 1, Postfach 1260, West Germany

H. F. KRAYBILL, National Cancer Institute, 9000 Rockville Pike, Bethesda, Maryland 20014

FRIEDHELM KORTE, Institut für Chemie der Technischen Universität München, D8051 Freising-Weihenstephan, and Institut für Okologische Chemie der Gesellschaft für Strahlen- und Umweltforschung mbH, Attaching, West Germany

WILLIAM C. KRUMREI, The Procter and Gamble Company, Ivorydale Technical Center, P.O. Box 599, Cincinnati, Ohio 45201

WOLFGANG KÜHN, Universität Karlsruhe, 75 Karlsruhe, Stolperstr. 5, West Germany

A. R. M. LAFONTAINE, Ministere de la Sante Publique, 14 rue J. Vytsmann, B 1050 Brusselles, Belgium

HAROLD H. LEICH, Drinking Water Coalition, 5606 Vernon Place, Bethesda, Maryland 20034

GEORGE H. LESSER, Environment Reporter, 1231 25th Street, N.W., Washington, D.C. 20037

DALE R. LINDSAY, National Center for Toxicological Research, Jefferson, Arkansas 72079

GORDON J. F. MACDONALD, Chairman, Commission on Natural Resources, National Research Council, National Academy of Sciences, 2101 Constitution Avenue, Washington, D.C. 20037

MAURICE MAROIS, Institut de la Vie, 89 Boulevard Saint Michel, Paris, France

THOMAS A. McCONOMY, Calgon Adsorption Systems, P.O. Box 1346, Pittsburgh, Pennsylvania 15230

J. K. MIETTINEN, Department of Radiochemistry, University of Helsinki, Helsinki, Finland

G. WADE MILLER, Public Technology, Inc., 1140 Connecticut Avenue, N.W., Washington, D.C. 20036

EMIL MRAK, Chancellor Emeritus, University of California, Davis, California 95616

SHELDON MURPHY, Harvard School of Public Health, 665 Huntington Avenue, Boston, Massachusetts 02115

S. FRED NOONAN, Bechtel Corporation, 1620 Eye Street, N.W., Washington, D.C. 20006

MARIAN PARKS, Environment Forum, 3021 Cambridge Place, N.W., Washington, D.C. 20007

RUTH PATRICK, Academy of Natural Sciences, 19th Street and The Parkway, Philadelphia, Pennsylvania 19103

FRANZ PERZL, Gesellschaft für Strahlen- und Umweltforschung mbH, 8042 Neuherberg, Ingolstadter Landstr. 1, West Germany

JOHN R. PITTMAN, Bechtel Corporation, 1620 Eye Street, N.W., Washington, D.C. 20006

RIP G. RICE, International Ozone Institute, 1629 K Street, N.W., Washington, D.C. 20418

THOMAS W. RILEY, Pacific Gas and Electric Company, 77 Beale Street, San Francisco, California 94106

GERARD ROHLICH, Chairman—Safe Drinking Water Committee, University of Texas, Austin, Texas 78712

THEODORE M. SCHAD, National Academy of Sciences, 2101 Constitution Avenue, N.W., Washington, D.C. 20418

LOUIS SIRICO, Citizens' Drinking Water Coalition, 1832 M Street, N.W., Washington, D.C. 20036

J. P. M. SMEETS, Health Protection Directorate, Commission of the European Communities, 23 Avenue Monterey, Luxembourg

EMMANUEL SOMERS, Director-General, Environmental Health Directorate, Department of National Health and Welfare, Ottawa, Ontario, Canada

MICHAEL J. SUESS, Regional Officer for Environmental Pollution Control, WHO Regional Office for Europe, 8 Scherfigsvej, 2100 Copenhagen, Denmark

WILSON K. TALLEY, Assistant Administrator for Research and Technology, US Environmental Protection Agency, 401 M Street, S.W., Washington, D.C. 20460

ANN TASSEFF, Editor—Environment Reporter, The Bureau of National Affairs, Inc., 1231 25th Street, N.W., Washington, D.C. 20037

RUDOLPH N. THUT, Weyerhaeuser Company, 1628 23rd Street, Longview, Washington 98632

A. RURIC TODD, Pacific Gas and Electric Company, 77 Beale Street, San Francisco, California 94106

RENE TRUHAUT, Director, Laboratoire de Toxicologie et d'Hygiene, 4 l'Avenue de l'Observatoire, Paris, France

H. G. S. VAN RAALTE, Shell International Research, Maatschappij B.V., Toxicology Division, The Hague, Netherlands

JOHN WELCH, A/C Pipe Producers Association, 1875 Connecticut Avenue, N.W., Washington, D.C. 20009

Introduction to the Series

Sponsored by the International Academy of Environmental Safety (IAES), this volume, "Water Quality," is the first in the new Series of publications. The need for a Series on "Ecotoxicology and Environmental Quality" is real, since most of the papers and/or review articles dealing with the problems of the environment are contained in journals. The new Series of publications gives the Editors the opportunity to publish not only proceedings of meetings of value to the scientific literature, but it enables them to present extensive review articles.

In conjunction with the Series, a new International Journal, called *Ecotoxicology and Environmental Safety*, will be published in the near future, also sponsored by the Academy. The Editors will coordinate the contents of both publications dealing with the problems of the environment, and particularly those relating to the ecological aspects of toxicology.

The Editors are cognizant of their responsibility. They will do all that is possible to make certain that these two publications fulfill the needs of the rapidly advancing frontiers of Ecotoxicology and Environmental Safety and Quality.

OPENING REMARKS

DR. COULSTON: The International Water Quality Forum is sponsored by the International Academy of Environmental Safety (IAES) and the Environmental and Agricultural Foundation, and it is supported through a grant from the Environmental Protection Agency (Grant No. R-803954010).

We will discuss more about the purpose of this meeting in a few moments, but I would like, at this time, to take the opportunity to introduce our host for this meeting, Dr. Thomas Kimball. Dr. Kimball is a well-known conservationist, and through his leadership, the National Wildlife Federation has grown to represent three and a half million people in the United States of America. I understand that in great part, he deserves this honor and distinction for developing the Federation. I think it goes without saying, that I could talk for over half an hour about his accomplishments, but I think it best to say only how pleased we are to be here, to have a meeting of this kind, in a hall represented by his Federation. It gives me great pleasure to introduce to you, Dr. Thomas Kimball.

DR. KIMBALL: Thank you, very much. It is my pleasure to welcome you to the National Wildlife Federation's Hall of Fame. You can see by the pictures surrounding you here, that there are a great many people who have had an impact on environmental affairs in our history, and you will see their names and a little bit about them depicted on our walls of this particular room.

For those of you who may not know, the National Wildlife Federation is a citizens' organization. Although we have a staff of some 400 people, representing many professional disciplines in the field of wildlife and environment, we are essentially a grass-roots organization, having citizen membership in all fifty states, and the three territories. The membership indicates the interest of the citizen in the quality of life associated with our natural resources, particularly the wildlife resource. So, we are particularly pleased to welcome this distinguished group of scientists here to address one of our critical issues; mainly, the quality of our nation's drinking water supply. Just as an indication of our interest in this particular subject, in August, we hosted a meeting in this very room of the Citizens' Drinking Water Coalition, which is a coalition of environmental and public interest groups throughout our country that are concerned about the implementation of the Safe Drinking Water Act, which was passed in December of 1974. And, it held its first conference here in this room, and addressed some of the same issues I am sure you will be discussing today.

So, we would like to welcome you. We hope that your deliberations will be meaningful and helpful in bringing to our nation the reputation that it has and hopes to continue in having safe drinking water for our citizens. Anything that our staff can do to make this particular meeting successful, please call on us.

I will close by saying that I will be talking to you briefly tomorrow, so I will reserve any further comments until then. Once again, welcome, and best wishes for a successful meeting.

DR. COULSTON: We thank you, very much, Dr. Kimball, for all of us, and we hope we use your house well and wisely.

A few statements that I think are important. The active working group of the meeting is the group at the front of the meeting hall. These are the so-called, forgive me, experts, that have been brought in from around the world and the nation. The observers, or participants, are sitting towards the back of the room. People with nameplates are the actual participants of the meeting.

All observers, as they come and go through the next three days, are free to say anything they wish, and hopefully, we will have a discussion between the working group of experts and other experts in the room. Please hold up your hand when you want a microphone, and identify yourself, and the chair will recognize you, then there will be no confusion. We are taping most of the discussions, as well as the presentations, and we hope to put this forum out in published form.

Dr. Werner Klein, from Germany, has consented to act as the secretary of the meeting. If you have any questions, or if you have said something you want deleted from the tape, please feel free to tell him. If you wish, simply state that something is not for the record, and this will be sufficient to keep it off the record.

Now, I have the privilege to open the meeting and as such, it gives me great pleasure to introduce the Chairman of the meeting this morning. He is an emminent and distinguished professor, scientist and scholar in France. I have personally known this man for almost twenty years, and many of us around the front table have sat on committees which he has chaired in the World Health Organization (WHO), Food and Agricultural Organization (FAO), International Atomic Energy Agency (IAEA) and others. He has, in one way or another, been involved with international meetings concerned with a cardinal principal - the safety evaluation of chemicals and drugs as they affect the people, the animals, and the plants of the world. So, we are very fortunate to have with us, the man who will give the opening remarks. He is, at the present time, the President of the International Academy of Environmental Safety, a distinguished career scientist and a gentleman. It gives me great pleasure to introduce, Professor Dr. Rene Truhaut.

DR. TRUHAUT: Professor Fred Coulston, I turn first to you, and I hope, Ladies and Gentlemen, dear colleagues, that you will forgive me in saying to my friend that he gave to me so many flowers that I am almost surprised to be still alive. Thank you, very much. These words came from your heart, and effectively we are friends and colleagues for so many years, and we work together objectively with good will, and thank you, very much for what you said.

Ladies and Gentlemen, it is a great honor for me to present at the beginning of this first session of this forum, the opening remarks.

The reason for this honor is, as Professor Coulston mentioned already, that for the moment, I am the President of the International Academy of Environmental Safety, which is one of the organizations sponsoring this forum. I am the President, having succeeded Professor Coulston, who was the first President of the Academy. And, according to the rules, we have to be President for two years. My term of office will end this year, and at this moment, the President-elect will take the chairmanship. That is Dr. Emil Mrak, Chancellor Emeritus of the University of California at Davis.

And now, this is a nice opportunity to thank both of them and, in addition, to thank the man who is sitting between them. This man is Professor Friedhelm Korte, the General-Secretary of the Academy, and these three men took care to organize this meeting and to plan the program. I think it is only justice to express our gratitude for the job they have done and for the remarkable manner they have planned the program.

I have also the privilege to say some expression of gratitude, not only as Professor Coulston made to our host, Dr. Thomas Kimball, but also to the Environmental Protection Agency, which gave us the grant for organizing the forum. And, I also think, to all those who took care of preparing the meeting and helping the three men I have already mentioned.

Ladies and Gentlemen, you have already realized, in listening to me, that I do not speak fluent English. I might have said, in the beginning, that I was born in Texas, but my friends said to speak frankly, and for this reason, I would say that I was born in a rural part of France. I hope you will forgive me for my heavy accent, and bear with me during my talk, and I further hope that you will understand the main ideas that I would like to express.

I feel I should give some information about our Academy. Our Academy is, I repeat, one of the organizations sponsoring this meeting. The other organization being the Environmental and Agricultural Foundation, which is located in California, and of which Dr. Mrak is the President. And here, again, I have a duty to express our gratitude. I must also welcome all of you who came from abroad, because if I look around the room, I see faces I know already. I see others I do not know, but I have to stress that you are coming from many countries, of course from the United States and Canada, but also from Western Europe, Finland, from Japan, from the United Kingdom, Luxembourg, Belgium, Netherlands, Italy, Switzerland, and so on. And I repeat, it is a pleasure today to welcome you warmly, not only on my behalf, but on the behalf of all the members of the Academy.

To deal with the task I have to do this morning, I will discuss briefly the following items: the formation of the Academy, its aim, its composition, its achievements, and, briefly, its perspectives of activities for the future.

The International Academy of Environmental Safety (IAES) was founded on the 28th day of May in 1971 in Munich at the occasion of the Second International Symposium on Chemical and Toxicological Aspects of Environmental Quality. Some bold fellows met together and decided to found this Academy. Among them were Professor Frederick Coulston, Professor Friedhelm Korte, Dr. van Raalte and myself, all present at this meeting. What we had in mind, and now I am referring to the aim of the Academy, is the following:

As you know, we are living in a chemical era, resulting in diffusion of chemicals into the environment. As a result, these chemicals can be hazardous for the health of man. Also, more widely, for the well-being of living organisms in the environment. Plants, animals, micro-organisms, all are involved in this, and this is the basis for the new concept of ecotoxicology. I say a word about ecotoxicology later.

Looking at the current perspectives, we were obliged to observe that if some people have a tendency to ignore the hazards, there are others which, unfortunately, have a tendency to exaggerate these hazards. I will refer to the famous three monkeys, that is, "Hear No Evil, Speak No Evil, and Listen to No Evil!" Many of the people who are speaking about the environment are really very emotional, and instead of being scientific and objective, are stressing to the people that they are going to be poisoned. This is the reason why I refer to the three monkeys. These emotional people say "Drink No Evil, Eat No Evil, Inspire No Evil," that is to die and go immediately to Paradise. And this is, you will realize, a little exaggerated. For this reason, we decided to found this Academy, to try to study objectively and scientifically the hazards and to put in perspective, not only these hazards, but also the benefits for the welfare of mankind. And I hope that my friend, Dr. Coulston, will agree to this synopsis about the objectives of the Academy.

Now, the composition of the Academy. The Council consists of nine members: the President; the Past-President, Professor Coulston, who by the way, was nominated Honorary Chairman and a Supplementary Member of the Council (it is the least we can do to honor him, because he was really the pioneer in the foundation of the Academy); the President-Elect; the Secretary; the Treasurer; and up to four other members.

The number of members, we decided, is limited to 100, and for the time being, we have already 80. Which means, that we are to nominate 20 supplementary members. We hope very strongly that for the future, it will be through the cooperation of many scientists around the world that the aim of the Academy will be fulfilled. Two weeks ago, in a special session of the Council, we decided to extend the number of people involved without changing our statute. We decided to create a new international society, attached to the Academy, which is called the Society of Eco-Toxicology and Environmental Safety. Eco-toxicology, I have already referred to it, is a new branch of toxicology which has to deal not only with human health, but with the study of harmful effects on all the constituents of so-called ecosystems: plants, animals and micro-organisms. You know enough, all of you, about the example of the effect of mercury on fish, and after all, on man, and the effects of fluorides on plants, etc. The creation of the new Society I see as an achievement which will give the opportunity for many scientists around the world to join indirectly the Academy, and to give to us the benefits of their competence and their cumulative experience.

Now, about the accomplishments. Briefly, to give you an idea of our achievements, I will tell you we have held five successive meetings of the Academy on problems relating to the environment and human health.

In August, 1973, we met at the occassion of the conference of IUPAC (International Union of Pure and Applied Chemistry). We had the pleasure to have with us at that time, the President of the Applied Chemistry division, Dr. Harold Egan, from the United Kingdom, and it is a pleasure to welcome him especially, and to tell him we are very pleased to have very close connection with IUPAC, because it is impossible to study environmental problems without paying attention to analytical problems. You cannot, for example, establish dose-effect relationships without having at the same time the possibility to study the environmental effects and to determine the degree of exposure. And, in this way, we are able to satisfy the Golden Rule of Pharmacology and Toxicology, which was written centuries ago by our Father, the Great Paracelsus. If something is toxic for some people, it appears to be toxic for all; this is not true. Otherwise, we would not be able to drink any alcohol or wine, because alcohol is really a toxic substance. Fortunately, its toxicity depends on the dose.

The last (most recent) meeting we held was in Munich, two weeks ago, about Ecological and Toxicological Aspects of Organochlorine Compounds. This meeting considered the scientific basis for the establishment of threshhold levels for carcinogens, and dose-response relationships.

Now to the present meeting. I see you, I see your faces. All of you are very enthusiastic. I see your eyes, you are very efficient people, and I think that this meeting will be a success! I wish all of us a good success with this meeting, I wish also that this meeting will reflect not only the efficiency and the alive character of the Academy, but also its tendency to keep in close liason with other organizations, not to overlap with them. It is the reason why it is sponsored also by another organization and supported by EPA. I will say, at the end, Ladies and Gentlemen - We have in mind not to found empires or to invade the field of other organizations, but we are in mind to cooperate with them.

DR. COULSTON: Thank you, very much, Dr. Truhaut. I would just like to add that the main purpose of the Academy is to present the ideas of a group of experts that can put together with unimpeachable, expert authority, arguments against some of the statements made by others that may not be scientifically correct. We are indeed an international organization, representing 38 countries with outstanding scientists, lawyers, politicians, philosophers as members of our Academy.

It gives me great pleasure at this point, to introduce the man who made this meeting possible, and a man we all look to with great respect and admiration. I don't have to go into his credentials, because everybody knows Dr. Emil Mrak.

DR. MRAK: Thank you, Dr. Coulston. The idea of this meeting started in France, about two or three years ago. When Professor Truhaut said we have problems on water, we ought to have a forum on water someplace, why not the United States? Because you seem very worried about water, especially drinking water. So, as a result, this meeting came about. It is not sponsored, but supported by the EPA, you can see the difference there. There is a lot going on in water here. My task is to introduce some of the people who are here, and I am only going to call on two: Dr. Gordon MacDonald, Chairman on the Commission of Natural Resources of the National Research Council, National Academy of Sciences; and, Mr. Ted Schad, who is up to his neck in water at the National Research Council. With that, I think I will turn it back to you, Dr. Coulston.

DR. COULSTON: I now turn the chair over to Dr. Truhaut.

DR. TRUHAUT: Thank you. We will now proceed, and I will ask you, Professor Coulston, to give your presentation about the Purpose and Significance of the International Water Quality Forum.

DR. COULSTON: It is, as you all know, always important to give the reasons why something is being done, so that strangers will understand that there is no vested interest, there is no motivation except to try and bring to the United States an International Forum on Drinking Water Quality.

PURPOSE AND SIGNIFICANCE OF INTERNATIONAL WATER QUALITY FORUM

Frederick Coulston

Water is universal, water is all around us. It is in our food and we cook with it. We drink it and we mix things with it. All body functions depend on water; and plants and animals learn to adapt their body functions to humid and dry conditions of their environment. We are water - at least 80% of our body is water with chemicals and physical agents dissolved, suspended or dispersed colloidally in the body water. Tissues of the bodies of plants and animals are designed to function - some to keep the water in - some to excrete it and each living thing has learned to recycle water within its organism in order to survive.

But - what is water?

The various countries of the world look to the USA and to each other for knowledge to provide adequate water of some mysterious quality for drinking, for agriculture, for industry, for recreational use and natural purposes. We feel impelled to have adequate supplies and a good quality of water.

But - what quality?

The third world countries cannot afford the research and regulatory agencies necessary to provide safe, adequate quality water. Therefore, they turn to us of the sophisticated developed world for guidance, advice, and general knowledge relating to good quality water. How good? What should be its composition? How safe? Do we need to take everything out? Should we use distilled, demineralized, charcoal filtered, and bacteriologically sterile water? What is water? What is water for the USA, India, Germany, Yugoslavia, Russia, Canada, Nigeria, Burma, Japan or Argentina and Venezuela? Let there be no doubt - eventually we all share the same H_2O!

On the one hand, we add poisons to our drinking water, such as chlorine and fluoride, and on the other hand, we talk about the poisons naturally present in water or acquired through human needs in body functions, agriculture or in industrial processes and manufacturing. We glibly discuss taking all chemicals out of water, if there is any suspicion that these may be carcinogenic, mutagenic or teratogenic (as if that is possible).

Yet, in India, whole villages must be abandoned because there is an excess of fluoride in wells used to provide drinking water. We, in our sophisticated society, relate the benefit of adding fluoride to prevent dental caries, against the toxic effect of the chemical on the nervous system.

The safety factor is so small between the effective and the toxic dose that fluoride could never be added to food in the same sense as an ordinary GRAS substance, such as sugar, salt, food colors and flavors, or food additives in general. Yet, we have selected a dose of fluoride that we say is safe for children - and commit adults to this dose for their lifespan, even when it is known that there is no benefit to adults. Let me hasten to say, I am not against fluoridization of drinking water resources. I am merely pointing out that what is in water should be considered at least in part on a benefit-risk and socio-economic basis.

We went into treating water with fluoride with our eyes wide open after a decision-making process. Yet, in India, people become sick and die when fluoride is in drinking water in excess amounts. At least we allowed the people in the water districts of the USA to decide for or against fluoridization.

Since many countries look to all of us for guidance in food and drug and environmental matters, it seemed fitting to the organizers of this meeting to hear first-hand from experts around the world, what are their water quality problems. Each country has different problems - and we must and can learn from each other. Therefore, this meeting is planned to discuss drinking water quality from the viewpoint of an International Forum. All nations must be involved eventually to establish an International Scientific Basis of Water Quality Standards. All of us share the same water. We must decide now how to preserve water resources and learn how to use and provide water to the advantage of all. That, then, it the Purpose of this meeting.

WHAT IS WATER?

Friedhelm Korte

Water, representing the medium of life on earth and one of the four ancient "elements", is a summarizing and general term, for which no exact scientific definition can be given. We do not know exactly where the huge amounts of water on the earth come from - but according to our present understanding of the evolution of the earth, the chemical H_2O must have been there from the very beginning.

Depending on the aspects and viewpoints of different approaches, the term water includes a large variety of different liquids. For all, there is only one common characteristic, namely, that the major constituent of water in any case is H_2O.

At the bottom left of Figure 1, you will see a hydrated proton - H_3O^+ - with its proposed structure. The molecules adjacent to the H_3O^+ ion have very high dissociation enthalpies resulting in a $H_9O_4{}^+$-complex, whereas the outer water molecules are less strongly bound. However, the life span of the single H_3O^+ is short (10^{-12} seconds) and proton transfer from one H_2O to the other occurs so quickly that the positive charge is delocalized along chains of hydrogen-bonded water molecules, resulting in an electron distribution comparable to conjugated systems.

The characteristics of hydrogen bonds even might be involved in the formation of mutations. However, I will not discuss this theory in detail.

Figure 2 shows that bulky structures of liquid water are a further consequence of hydrogen bonds and result in the well-known anomalies of water - the maximum density at 4°C, the already mentioned high values for evaporation enthalpy, high surface tension, and specific heat.

The correlation of density with temperature is shown as one example in Figure 2. As with other chemicals, there exist different structures depending on pressure and temperature; Figure 2 shows the phase diagram of the ice - liquid and fluid water system at pressures up to 200,000 kp/cm^2 with six different crystalline ice modifications. It shows that by increasing the pressure, liquid water can exist down to minus 20°C.

Like other chemicals, H_2O does not exist anywhere in absolutely pure form, and there were and are still great efforts to prepare water samples with decreased electrical conductivity, being a measure for purity. Apart from dissolved other chemicals which are the major topic of this forum, the chemically ideally pure normal water is not H_2O. It contains normally 150 ppm of heavy water, and following the natural oxygen isotopes, ^{18}O and ^{17}O-water. The further possible combinations - e.g. $D_2{}^{18}O$ - are insignificant.

Figure 3 shows some physical properties of H_2O and D_2O and major isotope abundances.

The general chemical properties do not differ between these water molecules. However, there are some more sophisticated differences in their behavior; I would like to mention as example the decrease of metabolic activity of fish living in D_2O enriched water. Differences in natural isotope ratios which are used in many research areas should be mentioned here for deuterium; differences in the deuterium content are measured as delta-D values, that is, the relative deviation from Standard Mean Ocean Water (SMOW). Investigations of ground water profiles, for example, showed that there is practically no exchange or mixing in deep layers and allow to conclude about the climatic conditions at the time of their formation.

Although chemical and physico-chemical properties of the pure chemical water are fascinating, I turn now to the aspects of this meeting: What is Water? under the points of view of its use or consumption?

For the agronomist, water is necessary to grow crops. He uses soil moisture, rainfall, and since natural water supply differs very much in climatic regions, he has frequently to use irrigation water.

Figure 4 shows some maximum levels for elements in water meeting the needs as irrigation water.

For the industrial engineer, the major use of water is as a cooling agent or as steam.

For the biologist, water is the medium of life, inevitably linked with life as we know it on earth.

For the meteorologist, water (rain) is a factor of climate and an energy stock to equalize fluctuations in temperature.

In nature conservancy, surface waters are the habitat of wildlife, and for the general population recreational aspects are important apart from drinking water.

According to viewpoints and needs of these groups given as examples, water has to have different qualities, that means different chemical compositions.

Figure 5 shows the global distribution of the waters of the hydrosphere. Of this huge amount of 1.36 times 10^9 km^3, man can use far less than 1% - in principle, half of the ground water, soil humidity and surface waters. Due to the uneven global distribution of these sources - e.g. half of the world's rainfall reaches the Atlantic Ocean as rainfall and from rivers - the amounts which are practically used are even far less. Figure 6 gives as an example the water supply and use in the German Federal Republic during the last decade. The high amounts used by industry are mainly for cooling purposes, uses for other purposes equal domestic and manufacturing use.

As regards supply of waters for different purposes, a major point of concern is the continuous increase of the consumption making recycling principally impossible.

Coming to the chemical composition of water, Figure 7 shows the content of some ions in rivers, the variation even for these major salts and the global average. As regards organic natural chemicals, amounts in the 1000 ppm range can be present in surface waters, especially swamps. Additionally to dissolved chemicals, particles of organic and inorganic structure are natural water constituents (Figure 8).

I would like to discuss, however, the examples of ocean water - one problem which is frequently misunderstood: accumulation or biomagnification. Figure 9 shows the biomagnification of chemicals present in ocean water in the copepode Calanus finmarchicus showing that specific accumulation of the elements from the environment of an organism is a typical function of life. In the framework of pollution, accumulation is discussed frequently together with water solubility or lipophilic properties of the respective chemical. However, it should be considered that organisms also accumulate water soluble chemicals like iron - ions and that on the other hand, water dissolves nonpolar compounds resulting in hydrophobic bonds of the nonpolar chemical - e.g. a hydrocarbon - in the structure of liquid water. Consequently, accumulation in organisms and food webbs is a complex interaction of the respective solubilities, resorption, biosyntheses or excretion mechanisms. There should be no principal difference in this respect for natural and man-made chemicals.

However, for those xenobiotics for which no effective elimination mechanisms exist in the living organisms, an unintended storage to certain steady-state levels may occur. Figure 10 shows the global mobilization and production level of some base products demonstrating the extent of potential anthropogenic changes of the composition of the global environment. For several elements, especially heavy metals, mobilization by mining exceeds geological enudation. Therefore, these inorganic ions were of major concern during the last decades and their impact on water quality has been thoroughly studied. A number of questions and problems remain to be solved like the mechanisms of formation and breakdown of methylmercury or the steady increase in drinking water nitrate levels which happens, for instance, in many places in Israel due to the need of frequent recycling the available water.

Today, our concern as regards the composition of water accessible and used by man should be directed in some way to organic man-made chemicals. In this slide, the production of organic chemicals is given with 70 million tons in 1971 - there are estimates that now the production is 130 to 150 million tons.

Figure 11 shows a calculation of the potential global concentration for the total of organic chemicals resulting from one year's production. One hundred million tons produced, released in some chemical form into one medium of the environment only, not degraded and dispersed homogeneously, would result in either of the concentrations given in the Figure - e.g. 0.3 ppm in the top 1m layer of the world oceans. This calculation might seem unrealistic - it is not an estimate - since there is degradation in the environment, waste incineration, and there is no short-term homogeneous dispersion. However, it shows that man is able to change the composition of the global environment with respect to synthetic organic chemicals.

A forecast based on present levels might be more realistic. Under the assumption that a linear correlation exists between input of chemicals in the environment and breakdown, high levels of organic chemicals might be expected. Assuming a 5% annual increase in the production of a chemical which is today present in a concentration of e.g. 1 ppb in a lake will result in 13 ppm of that chemical in 100 years from now; a growth rate of only one percent more - 6% - would result in 87 ppm of the chemical. This table demonstrates that we have to expect increasing amounts of organic chemicals in the environment and that we should consider this fact carefully (Figure 12).

When considering the classes of chemicals which might become problems, the Eurocop-Cost-List of chemicals identified in different waters shows individuals of all important classes of industrial chemicals apart from natural products. Figure 13 shows the major classes of chemicals summarized in this study, the most frequently found chemical compound and the number of individuals of each class. The high number of organohalogens might be due to the persistence of the carbon-halogen bond but might also reflect the excellent analytical procedures as compared to some other classes.

When looking at the number of organic chemicals identified in waters - or at summarizing parameters like organic bound carbon which will be discussed in another paper - and when remembering the factors given in the slide before, we can imagine what water might become in the future (Figure 14).

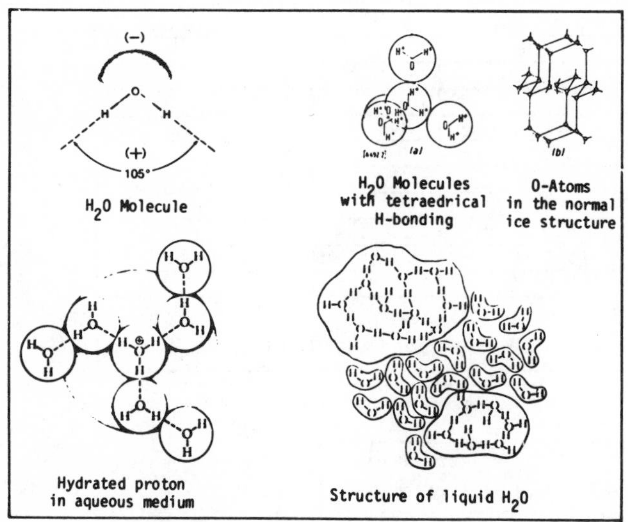

H_2O Molecule

H_2O Molecules with tetraedrical H-bonding

O-Atoms in the normal ice structure

Hydrated proton in aqueous medium

Structure of liquid H_2O

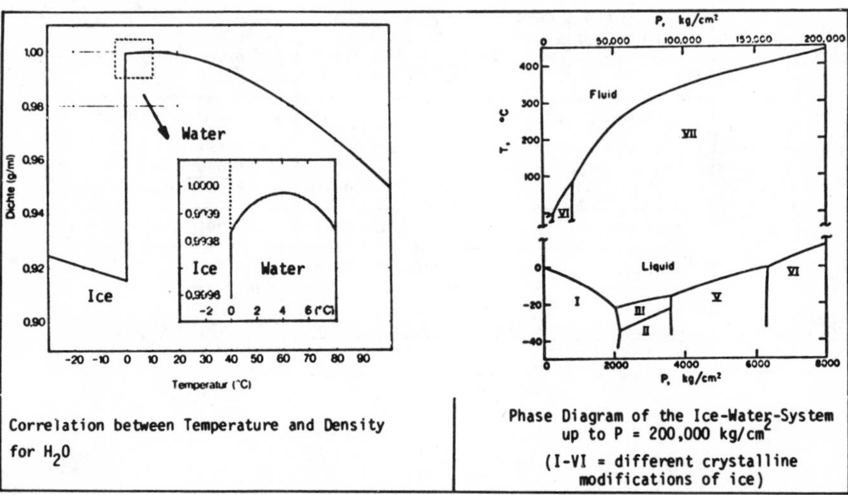

Correlation between Temperature and Density for H_2O

Phase Diagram of the Ice-Water-System up to P = 200,000 kg/cm^2
(I-VI = different crystalline modifications of ice)

	H_2O	D_2O		
density at 20° C	0.9982	1.1059	$H_2{}^{16}O$	100 000
melting point (°C)	0.00	3.82	$H_2{}^{17}O$	37
boiling point (°C)	100.00	101.42	$H_2{}^{18}O$	204
			$D_2{}^{16}O$	15

(Schröder, 1970)

LIGHT AND HEAVY WATER; Isotope abundances

element	for continuous use, all soils	for short term use on fine textured soils only
Cd	0.005	0.05
Co	0.2	10.0
Cr	5.0	20.0
Cu	0.2	5.0
Mn	2.0	20.0
Mo	0.005	0.05
Ni	0.5	2.0
Pb	5.0	20.0
V	10.0	10.0
Zn	5.0	10.0

(FWPCA, USDI)

TOLERANCES FOR IRRIGATION WATERS (in mg/l)

	global amount km^3	percentage of total
Oceans	1 321 890 000	97.2
Polar ice and glaciers	29 190 000	2.15
Ground waters		
up to 800 m depth	4 170 000	0.31
below 800 m depth	4 170 000	0.31
soil humidity	67 000	0.005
	8 407 000	0.625
Surface waters		
inland lakes	229 000	0.017
rivers and streams	1 000	0.0001
	230 000	0.017
Atmosphere	13 000	0.001
total	1 360 000 000	100

(Leopold and Davis 1970)

GLOBAL DISTRIBUTION OF WATER SOURCES

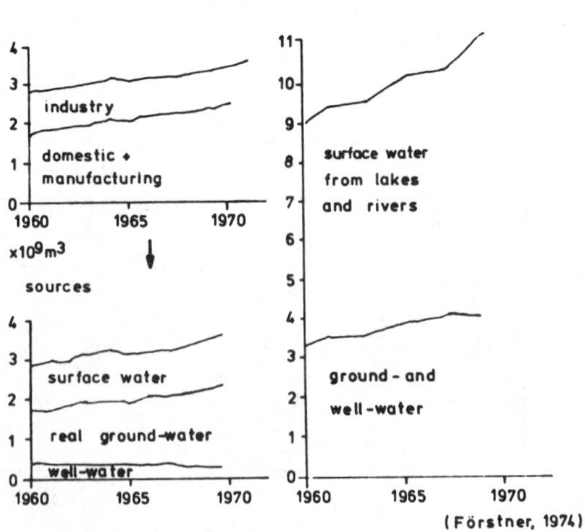

Water Supply and Sources in the Federal Republic of Germany

	Orinoco	Rio Grande	global average
HCO_3^-	22.0	183	58.4
SO_4^{--}	8.8	238	11.2
Cl^-	2.0	171	7.8
Ca^{++}	3.2	109	15.0
Mg^{++}	0.5	24	4.1
Na^+	8.7	117	6.3
K^+		7	2.3
Silicic acid	8.0	30	13.1
total	53.2	879	118.2

(Livingstone, 1963)

NATURAL SALT LOAD OF RIVERS (mg/l)

Element	Milligrams per liter	Element	Milligrams per liter	Element	Milligrams per liter
Chlorine	19,000	Zinc	0.01	Tungsten	1×10^{-4}
Sodium	10,600	Molybdenum	0.01	Germanium	1×10^{-4}
Magnesium	1,300	Selenium	0.004	Xenon	1×10^{-4}
Sulfur	900	Copper	0.003	Chromium	5×10^{-5}
Calcium	400	Arsenic	0.003	Beryllium	5×10^{-5}
Potassium	380	Tin	0.003	Scandium	4×10^{-5}
Bromine	65	Lead	0.003	Mercury	3×10^{-5}
Carbon	28	Uranium	0.003	Niobium	1×10^{-5}
Oxygen	8	Vanadium	0.002	Thallium	1×10^{-5}
Strontium	8	Manganese	0.002	Helium	5×10^{-6}
		Titanium	0.001	Gold	4×10^{-6}
Boron	4.8	Thorium	0.0007	Praseodymium	2×10^{-7}
Silicon	3.0	Cobalt	0.0005	Gadolinium	2×10^{-7}
Fluorine	1.3	Nickel	0.0005	Dysprosium	2×10^{-7}
Nitrogen	0.8	Gallium	0.0005	Erbium	2×10^{-7}
Argon	0.6	Cesium	0.0005	Ytterbium	2×10^{-7}
Lithium	0.2	Antimony	0.0005	Samarium	2×10^{-7}
		Cerium	0.0004	Holmium	8×10^{-8}
Rubidium	0.12	Yttrium	0.0003	Europium	4×10^{-8}
Phosphorus	0.07	Neon	0.0003	Thulium	4×10^{-8}
Iodine	0.05	Krypton	0.0003	Lutetium	4×10^{-8}
Barium	0.03	Lanthanum	0.0003	Radium	3×10^{-11}
Indium	0.02	Silver	0.0003	Protactinium	2×10^{-12}
Aluminum	0.01	Bismuth	0.0002	Radon	9×10^{-14}
Iron	0.01	Cadmium	0.0001		

ELEMENTS IN SEA WATER

	ocean water	Calanus finmarchicus	concentration factor
oxygen	85.966	79.99	0.93
hydrogen	10.726	10.21	0.95
chlorine	1.935	1.05	0.54
sodium	1.075	0.54	0.50
magnesium	0.130	0.03	0.23
sulfur	0.090	0.14	1.6
calcium	0.042	0.04	1.0
potassium	0.039	0.29	7.4
carbon	0.003	6.10	2 000
nitrogen	0.001	1.52	1 500
phosphorus	<0.0001	0.13	20 000
iron	<0.0001	0.007	1 500

Elements in ocean water as compared to Calanus finmarchicus (% weight, total 100)

		10^6 t
Organic base chemicals		70
Inorganic " "		250
Iron		400
Non-iron metals		22
Crude oil		2065
→ org. chemicals	100	
→ lubricants	20	

GLOBAL PRODUCTION OF INDUSTRIAL BASE-PRODUCTS IN 1971

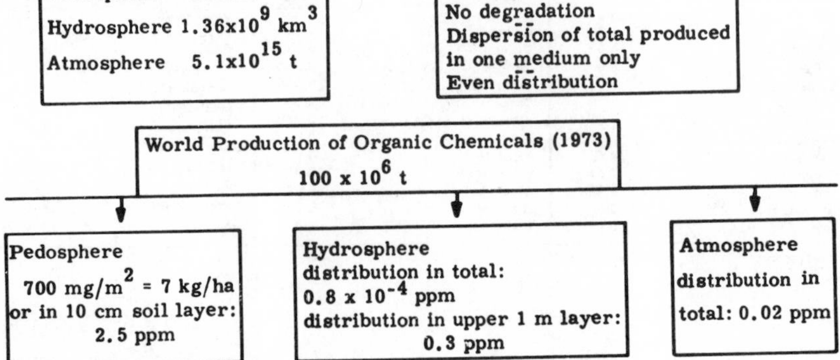

POTENTIAL ANNUAL LOADS WITH ORGANIC CHEMICALS

for	1% increase,	in 100 years			170-fold	present level		
for	2%	"	"	"	"	624-fold	"	"
for	3%	"	"	"	"	1 822-fold	"	"
"	4%	"	"	"	"	4 951-fold	"	"
"	5%	"	"	"	"	13 050-fold	"	"
for	6%	"	"	"	"	86 666-fold	"	"
"	8%	"	"	"	"	219 875-fold	"	"
"	9%	"	"	"	"	552 821-fold	"	"
"	10%	"	"	"	"	1 377 935-fold	"	"

FORECAST OF ENVIRONMENTAL LEVELS OF CHEMICALS as related to annual increase of production

class of chemicals	most frequent identified individual	number of identified chemicals
polycyclic hydrocarbons	3,4-benzpyrene	42
amines and derivatives	acrylamide	38
nitriles and azo compounds	acrylnitrile	10
nitro- and nitroso compounds	o-nitrotoluene	30
organophosphates	tributylphosphate	10
organohalogens	DDT/DDE/DDD	112
organometallic compounds	methylmercury chloride	5
mercaptanes and other sulphur-containing compounds	2,4-dimethyl-diphenyl-sulfone	31
phenols, quinones	phenol	59
heterocycles	dibenzofurane	32
surfactants (mixtures)	alkylbenzene sulfonates	7
fluorescent whitening agents (mixtures)	type p-aminostilbene	-
ethers, aldehydes, ketones	diphenylether	49
acids	palmitic acid	70
esters	dibutylphthalate	23
alcohols	2-ethyl-hexanol, α-terpineol	33
arylalkanes	ethylbenzene, styrene	48
alkanes, alkenes	hexadecane	37
amino acids, proteins	alanine, glutamic acid	20
carbohydrates	sucrose	16

(Europ-Cost 64 b, 1974)

ORGANIC CHEMICALS IN WATERS
(surface, waste, leaching of land-fills, ground- and drinking water)

compound	Concentrations in ng/l (ppt) Lake of Zürich surface	Lake of Zürich 30 m depth	well water	ground water	tap water
benzene	28	22	18	45	36
trichloroethylene	38	65	5	80	105
tetrachloroethylene	140	420	12	1850	2100
chlorobenzene	3	12	-	14	6
dichlorobenzene	16	26	-	-	4
trichlorobenzene	6	42	-	-	4
camphor	12	2	2	-	2
naphthalene	8	52	-	-	8
pentadecane	16	4	2	2	4
heptadecane	20	4	1	-	3
diphenylether	48	8	-	3	3
tributylphosphate	82	54	-	10	14
gasoline	~2000	~100	~50	~800	~800

(Grob, 1974)

EXAMPLES FOR ORGANIC CHEMICALS IN WATERS

ENVIRONMENTAL WATER RESEARCH

Wilson K. Talley

When I was approached about this conference, I said, "Good - I can find out from you people what I should do." Thus, I would propose to spend most of my time listening to you, after I lay some ground work, rather than presenting a talk.

I am the Assistant Administrator for Research and Development of the Environmental Protection Agency, which is a regulatory agency. We conduct our research program to support the regulatory function; either to set standards or to develop the technology to implement the standards. We regulate individual pollutants and are concerned with the control of their effects. We follow an individual pollutant as it is transported and transformed in the environment, in the soil, in the air and in the water. We are really concerned with the impact of the pollutants on receptors; that is, if we are talking about man, we want to know how a particular pollutant gets into the body through the air, through the water, or through food. Legally, we are limited in our ability to regulate pollutants in food. The Food and Drug Administration has the responsibility for the food consumed by man and we only have peripheral responsibility for analyzing that entry. With respect to air regulation, we can split this area up into the ambient, the occupational and the indoor environment.

With respect to ambient air, we are pretty well covered. We can monitor and perform correlations to find out what the dose response is for pollutants that come through the general atmosphere, though not as well as we would like - but we are improving. In terms of the occupational exposure to air pollutants, once again, we run into a legal hangup in that the statutory responsibility for the occupational environment is with the Department of Labor, specifically, the Occupational Safety and Health Administration. Indoors, we are as yet unconstrained as to mandate, but we cannot adequately monitor the indoor environment. We do not have personal dosimeters yet, although within the last few months we seem to be making a great deal of progress towards indirect monitoring of atmospheric exposure in the indoor environment.

In terms of water, however, we have it all. We can regulate and perform the research to provide the basis for that regulation in the two major areas: in the treatment of water wastes, and in water supply.

Concerning work on the treatment of water waste, we developed our programs into three areas: municipal waste water treatment, that is, sewage treatment or the separation of storm effluents from sewage; industrial water treatment, where we are concerned with specific effluents, which we handle on an industry by industry basis; and non-point waste waters, that is, waste waters from sources that are distributed - feed lots, farms, etc. This kind of pollution primarily results from agricultural practices where we worry about salt intrusions, a particular pesticide, agricultural chemicals, feed lots, and the like.

With respect to water supply, the EPA - and before it, the Public Health Service - had the responsibility for water supplies involved in interstate commerce. This work involves the studying of standards of the water that you would get on a train, an airplane or, in the last ten to twelve years, on a Greyhound bus.

In December 1974, the Safe Drinking Water Act became law, completing our authority to handle the question of environmental pollution in waters. The Safe Drinking Water Act is a national, and not a Federal, program. The burden for providing the supply of safe drinking water for the citizens of the United States rests with the States and the local governments. We will have to render assistance to the States but, in the main, they will have to develop the resources to carry out the program.

That brings me to my problems as Assistant Administrator for Research and Development in EPA. It is simply a question of priorities. The first priority is one on which we do not have much flexibility and that is the most recent legislative act, the Safe Drinking Water Act. This Act compels us to provide assistance to the States, principally with respect to quality assurance programs. This means that when State, local and commercial laboratories measure what is in the water, it should be equivalent to the same measurement done elsewhere in the country. But then we come to the waste water problem.

Right now, as I pointed out, we have split our research in this area organizationally into municipal, industrial and non-point pollution. In the municipal area, the Nation is investing $18 billion dollars in federal monies and another $6 billion dollars in State and local monies in sewage treatment facilities. Yet I have been told that our major problem is not the municipal water supplies, when we finally get that corrected, but industrial contamination.

Furthermore, if and when we manage to get the municipal and industrial situations under control, we still won't have managed to solve the total problem - the real culprit is the non-point contamination. I have asked the National Academy of Sciences, which has a series of study panels taking a look at environmental research and the use by the Agency of such research, if it will verify that statement. If it turns out that we are correct in our assumption, then my question is, should we shift our priorities and begin to spend more of our Research and Development monies on the non-point sources, less on the industrial and finally, least on the municipal problem?

I have also asked, "Can we agree that the priorities be set so that we attack those problems that are critical, and very, very important?" Furthermore, should we put on our list of problems those which are solvable with a little effort and get them done? Solving these would build a good record for us; they do not take much in the way of resources and they would then be out of the way.

A third question is, should we attack those problems where we have a great leverage? That is, if our limited budget is matched two for one or even nine for one by some other source of money, and even more important, by manpower, should we attack those problems?

Another difficulty is the fact that our regulations are based on two criteria: one is effects and the other is technology. If we can demonstrate an adverse health effect, then we can regulate that pollutant; and in some areas, if we can demonstrate that it is technologically feasible to remove a particular pollutant, then it must be removed. The question then becomes, should the EPA's Research and Development budget be shifted to do research on technology, or to do research on effects?

While on the subject of effects, we recently have come to a point technologically where the chemists are able to detect concentrations at much, much lower levels than previously. That is, we now analyze to the parts per billion of chlorinated organics, but we do not have the health effects data to decide whether or not this is significant. The public is not being informed; it is being alarmed by reports coming out of my shop, and we are in a quandary. What is needed, of course, is research on the effects of long-term, low-level exposure, chronic exposure to trace contaminants, rather than research on the acute effects.

Further complicating this issue is the fact that there have been statements that only ten to thirty percent of cancer is due to genetic effects, which means that the rest must be due to environmental factors. Cancer, of course, is treated differently in the public eye than other toxic effects of pollutants.

I have three questions to pose to you. One is: What can be said about the establishment of research priorities in terms of emphasis given to technology versus effects research? What can be said about the priorities among research on municipal waste waters, research on industrial sources, and research on non-point sources of pollution? And finally, in the area of effects research, what is our best route to establishing the effects of long-term chronic exposure to low-level pollution?

WATER QUALITY - MANAGEMENT OR MISMANAGEMENT?

Gordon J. F. MacDonald

In the United States, the incidence of water-borne diseases first led the Federal government to promulgate guidelines with respect to water treatment for drinking purposes. Attempts to control insect vectors living on water bodies further involved the Federal government in such efforts as the malaria and yellow fever eradication programs. These activities of the late 19th century and early 20th century were followed with some major exceptions by almost fifty years of Federal neglect of water quality. Techniques to transform muddy, often smelly water into something the populace would drink as it came out of the tap evolved rapidly, however, and public acceptance of tap water may have been enhanced by the United States' -- fortunately brief -- experiment with Prohibition. During this period, human health was the prime consideration as far as water entering the distribution system was concerned. Little or no attention was devoted to what happened to the water as it wended its way to the consumer, perhaps in lead pipes, or more recently, in supposedly sterile polyvinyl chloride plastic tubings.

The questions of drinking water, its quality and how the total treatment delivery systems can be improved will be the major focus of this meeting. Therefore I will leave this subject to the experts and instead examine the United States' policy for achieving water quality. As defined by the Federal Water Pollution Control Act of 1972, Public Law 92-500, water is considered to be of acceptable quality if it is, and I paraphrase the relevant language, "swimmable and fishable". This goal, which is mandated to be achieved in U.S. waters by 1983, will require a massive expenditure of monies through a public works program for the construction of sewage treatment plants and further vast expenditures in the private sector to meet the effluent requirements promulgated by the Environmental Protection Agency pursuant to PL 92-500.

How did the United States develop these policies and why? Is the present program of financing municipal waste treatment plants by the Federal government, and regulating industry through a massive bureaucracy, the way to achieve the "swimmable-fishable" water that so many would wish? Why have we as a country decided that legislated regulations are preferable to a system that depends on the economics of the marketplace? These are all basic questions, not considered in any detail by those that framed the legislative straightjacket we now find ourselves in. Even the National Commission on Water Quality, with the mission of attempting to predict how well the Federal

Water Pollution Control Act can work, has in effect accepted the regulatory basis of PL 92-500, and plunged ahead, assuming that enough Federal bureaucrats can make bad legislation work.

The case that Public Law 92-500 is unworkable, I claim, is self-evident. Each industry or plant will require a permit for discharge into the waters. The permit requirements and conditions are prescribed by the regulators. To the novice, so far, this seems reasonable and workable. There are, however, more than 50,000 such permits at a minimum that must be administered and enforced. Some 800,000 to 1,000,000 reports will be required annually, according to the National Commission on Water Quality. The opportunities for variances, for delays, for appeals on the grounds of unemployments, are endless, and already have been employed in countless cases. No matter how many well-meaning and tireless servants EPA manages to recruit, the task is unmanageable. As an aside, not only is the administration of tens of thousands of detailed technical permits unworkable, but the administration of the thousands of permit watchers is also a herculean task, requiring administrative capabilities that reside neither in EPA nor, for that matter, in any other governmental regulatory agency.

The history of the U.S. Federal government's reliance on regulation, use of the courts and of legislated police power is a long one, influenced in no small part by the abundance of lawyers in positions of high importance. The Water Pollution Control Act of 1956, the Water Quality Act of 1965, the Clean Waters Restoration Act of 1966, the Environmental Quality Improvement Act of 1970, are just samples of how the Federal government has attempted to force water quality by applying the force of the Federal government. These acts vary in the sense of where detailed responsibility lies, but the message is clear: Use the police power to regulate. The 1972 Act brings it all together and says look at what comes out of the pipe and use the courts to stop it. I sometimes wonder if this means that District Court Judges must wield dippers into the effluent as I once saw former President Nixon sample the effluent of a Chicago tertiary treatment plant.

The results of the present approach have not produced convincing results. Rivers that were dirty five years ago are still dirty despite the claim of EPA statistics that there have been marginal improvements. Millions of gallons of raw or only slightly treated sewage flow into the waters daily. These waste waters are not only the result of combined sewers; rather they are the result of inadequate management, facilities, and upkeep of the waste treatment works. Further, any economic analysis of our current program can only conclude, and have, that a uniform regulatory system as mandated by Public Law 92-500 is inefficient and inequitable. We are spending far too much to see the quality of our water improve imperceptably.

Two concepts, at the heart of our current program, doom this program. The first is that of uniform national standards. What a kraft bleached pulp mill can discharge into the Connecticut River, and what it can discharge into the Savannah are very different if the goals of "swimmability and fishability" are to be achieved in the most efficient manner. The presence of other industries will, of course, influence any such judgment. The important point is that one cannot dismiss the regional characteristics of water quality. Any watershed is a unity that cannot be overlooked in the desire to achieve National Water Quality objectives.

Present legislation benignly neglected the basic regional character of water. The trend in this direction must be reversed. We cannot manage water in the Delaware River the same way we manage water in the Rio Grande. Science, so often neglected, tells us that these two systems are different in hydrology, climatology, erosion, and so forth. The character of the region through which these waters flow is, to put it mildly, somewhat different. The effluent requirements in both cases will require a detailed consideration of seasonal flows, of seasonal water requirements, of the public perception of the water uses; in short, of the ways in which these waterways can enrich the lives of those who are fortunate enough to live near them.

The second shortcoming in our current approach is the total commitment to regulation. Regulation has become the American way of attempting to solve problems. Regulation is, in principle, simple. Thou shalt not; and if thou does, then the dreaded police will be there to do you in. I claim this is utter and complete nonsense. Regulation of public goods, water, air and land, has not and will not work. Regulating requires a knowledge of what is to be regulated, how the regulations are to be enforced, and how to overcome pressures to keep the regulations from being subverted. I do not need to recall the history of such agencies as the Interstate Commerce Commission, the Federal Power Commission, the Food and Drug Administration, and so on, to suggest that we have not as a country achieved great success in regulation.

One alternative to regulation is the use of the power of economic persuasion. The concept of user charges has a long history and a number of articulate advocates. But, the concept has never received the attention of policy makers that I think it deserves. While serving in government, I worked hard at advocating the notion of an emission charge on an air pollutant, sulfur dioxide. Despite the great intellectual and practical appeal of this advanced concept, it failed to receive even a single Congressional hearing. With that background of success, it is clear that we cannot anticipate major advances in the use of water charges, particularly if I advocate them.

The principal argument for the use of user charges is that they seem to have worked when used. In the case of municipalities, user charges have succeeded in reducing waste discharges, not just retarding their growth. A typical and not extreme example is Cincinnati, which reported a 40 percent reduction of industrial waste after installing a small user charge.

The reason that user charges can be effective is simply that when a price is placed on waste discharged, the plant operator has the strongest possible incentive to find ways to reduce that discharge. With respect to regulations, he can deal with them by giving the complaint over to his trusty legal counsel. If his overall economic operation of the plant is at issue, then the operator will consider all sorts of options. How can the production process be optimized for economic productivity, inducing reduction of water user charges? The game is no longer one of getting away with a minimum fine, but rather how does one actually reduce production costs by reducing waste?

For one who is intensely committed to the working of the market system, free insofar as is possible from regulation, the imposition of environmental regulations represents nothing less than the further intrusion of the Federal government into the activities of the private sector. The private sector does make use of common goods, air, water, and land. But for the last, it has paid no price. We as a people, in this country, and elsewhere, must recognize that a price must be paid for those goods, used freely in the past. "The Polluter must pay" is not an empty slogan. That statement makes real the economics of a free, rather than a state-controlled economy. There is no better way of implementing the principle that the user should pay, than requiring those who would deposit their refuse into the waters that belong to everyone to pay for that privilege so that those who wish to restore those waters have the means to do so. In the longer term, the use of the market economy, the resulting transfer of information and the incentive to develop production methods that ensure water quality will achieve the goals of "swimmable and fishable" long before the bureaucrats in Washington unstaple their forms.

DISCUSSION

Dr. Talley, can you define non-point sources, and OMB, too?

DR. TALLEY: OMB is the Office of Management and Budget, and it is the President's fiscal watchdog, making certain that each agency's budget doesn't exceed bounds so that the total budget message to Congress is within the limits the President wishes. In doing this, they make the trade-off between apples and oranges and pears and camels. They don't do a bad job, but it's clear that some of us come up shorter than we would like.

Non-point sources are those sources that are distributed. a municipality collects its sewage and puts it all into a treatment plant and becomes a point discharger into a river or into aquafiers. Industry has a plant, it does its processing, it takes in water and it discharges water from a point. But a farmer irrigates his field and the water percolates down into the soil or runs off into a river. A feedlot may stretch out over hundreds of acres and there may be collections of feed lots.

I think the real distinction isn't in the geographic distribution of these things, but the fact that municipal and industrial waste waters are the results of the decisions of very few decision makers. Whereas the non-point sources are the culmination of the effects due to hundreds of thousands of individual decision. And the fact there is pollution coming in the waste waters is not even the second or third highest priority in the thinking of these decision makers.

QUESTION: Dr. Talley, What is your budget, and how much are you asking for?

DR. TALLEY: I will talk about the fiscal year 1976 budget because what I am going to ask for, I will not get. In fiscal year 1976, if the appropriations process moves through correctly, we will have about $250 million in new obligation authority. We do not split out budget up neatly into air, water, and the rest, although we do have about $40 million in health effects research. And, I would say, something like $16 million in ecological effects. We are probably spending roughly comparable amounts in each of the three areas of municipal, industrial, and non-point pollution control, or maybe a little bit more in the municipal than in the industrial and the non-point sources. So the question is, is that correct? Is there much to be gained from making major shifts in the way we spend the dollars, especially given the limited manpower resources we have. That is the other question that you should have asked - how many people do we have?

In 1976, we will have roughly 1756 permanent full-time employees. That turns out to be fewer people than the predecessor agency programs had in 1966, with about three times the money. The resources I speak of now refer only to EPA's Office of Research and Development. The Agency as a whole has about ten thousand people.

Someone mentioned the fact that you cannot produce the dose-response functions. Sticking with PCB, I think we can get the risk. The mink producers in the Great Lakes regions have known for some time that if you feed the fish from the Great Lakes to their mink, the mink will not breed.

It is quite easy to make economic calculations of the effect and the risks of using certain contaminants. What people fail to recognize, and now I come down on the strict environmentalists, is that at some level you may decide that the risk is negligible, especially after a great deal of dilution in water supplies.

The one thing that you cannot do is continue to use the concept of the threshhold. I think that we are talking in terms of public health. But, at some point, we are going to have to go to blind arithmetic and linearity and all sorts of assumptions, as has been done with the radiation limits that you mentioned, and know that we are going on supposition and guessing. But when we do that, I think we should be willing to admit that that is what we are doing, that there is little scientific validity to the threshhold concept beyond an urge to protect the public health and welfare.

QUESTION (Dr. Ernst): I was very glad to hear something about fluoridation from Professor Coulston, because I want to learn something about this problem in the United States. I have learned that there are new apparatus on the market to take the fluoride again out of the water. And I think that we have to recycle water, and only water. Ninety-eight percent of the water, anyway, is used for car washing and for restrooms, etc., and people could buy a fluoride toothpaste. But this is not what I am scared about. I am more scared about when people explain to me that it's a social-political problem and some parents cannot take care of providing toothpaste for their kids. But I think its the wrong way, and can be very dangerous for the future if we try to solve social-political problems with water. Because I think it is not far away that somebody will bring up the idea that we should dissolve the anti-baby pill in the water and give it to the consumer. I think that's the wrong way to make politics. And, therefore, I thank Professor Coulston that he brought it up.

DR. COULSTON: There is no answer necessary, of course. I used fluoride only as an example. You could use chlorine, or salt, as examples. The issue is, what do you put in the water and what do you take out? Is it necessary to take everything out of the water put in by our civilization in a so-called developed country? Do you have to take all the detergents out? Do you have to take out all of the color? Will it hurt man? The obvious answer is that a lot of these things will never hurt man. It is just a question of the economics and safety evaluation, whether you want your rivers flowing with nice bubbles from the detergents in them, or you want them without. (Frankly, it's very pretty to see a waterfall with bubbles being made by a little detergent in the water.)

The idea, though, is simply this. We must define, we must find out what is safe, what is harmful, what is economic to take out, what isn't economic to take out of the water. We must protect the jobs of the people. I agree very much with the last speaker. We have no right to deprive people of jobs in a local water shed community when they themselves will make the decision whether they want to live with a little bit of something in the water that may be harmless. If the Federal agencies say take it out, then the factories may be closed, like in the area where I live. Hundreds of people have been put out of work, because paper companies cannot economically remove certain waste products from their effluents and the factories have been closed.

With fluoride, you can decide on the basis of a decision making of the people of that area. I endorse this concept of Dr. MacDonald's, very much.

QUESTION (Dr. Junk): I would like to raise a point here, and possibly get Dr. Coulston to comment on it. He mentioned in his presentation that one must consider the benefit-risk ratio to decide whether we should add chemicals or attempt to remove chemicals from water. We also heard a talk this morning about non-point sources, and I will refer specifically to something like the pesticides and herbicides, where it is fairly easy to scientifically arrive at the benefits associated with the use of pesticides and herbicides. It is extremely difficult to arrive at the risk associated to humans through the use or ingestion of particular pesticides and herbicides. And, if one looks at it realistically and scientifically, I doubt if within the next 10-15 years, we are going to have an accurate assessment of the risk associated from the ingestion of pesticides. Therefore, it strikes me as if it is almost impossible to use the benefit-risk ratio in order to decide whether we ought not have a particular chemical in our water supply.

DR. COULSTON: I agree with what you say, but I don't think I meant it in the way that you are interpreting it. The benefit part would be to use the pesticide for agricultural or public health purposes for the benefit of the community and of the people.

Knock out the mosquitoes that transmit encephilitis, let's say, or the fleas in New Mexico. We now have some cases of Bubonic Plague. Eleven deaths so far this year. Bubonic Plague, in this year, in the United States, and they are trying to knock out fleas with pesticides that are no good for fleas. The fleas laugh at them. There are pesticides known that would knock the fleas out.

But, and the issue is, the benefit has to be clear, and then we can decide whether a residue is acceptable, or permissible. That is the issue, you see. The same is true with radiation. We can't argue that radiation particles are good for us. You may be able to set a threshhold level for expedience sake, but one could argue an atomic energy plant is absolutely necessary for a certain community where you can't get gas, oil or coal. And you have to depend on this type of fuel for steam production in order to make electricity. So, there you again weigh the benefit-risk.

But with benefit-risk, always must come a socio-economic decision. Most of our basic decision are really not made by scientists. They are made by lawyers, judges, and the sociologic side of our environment, if you will allow me to put it that way. I think there is nothing wrong with this. I don't think scientists should make these decisions alone. They should be listened to, but, in the final analysis, the lawmakers representing the people have to make the decisions. There is nothing wrong with that.

Dr. JUNK: My real point is that it is very difficult to put numbers on that risk.

DR. COULSTON: But you know, many of us sitting here have chaired meetings where this very process is carried out every year in Geneva, Switzerland, for the WHO/FAO. We do, indeed, sit and we do make these decisions on many chemicals, be they food additives or pesticide residues. I don't want to go into the elaborate way in which this is done. We set tolerances, we set ADI's (Approximate Daily Intake) of chemicals, etc.

MR. CLARK: I am Carl Clark of the Citizens' Drinking Water Coalition, a small group to inform the public of some of these problems. And, Dr. MacDonald, what is the procedure that you would recommend for monitoring these waste products? Would you have a regulation that industry has to report its own pollutants and then pay this level? Or could anybody monitor? It is still a bureaucratic problem.

DR. MacDONALD: There will always be a problem in the question of who and how monitoring is to be carried out. That problem, I think, has been over-emphasized as to its severity in many cases. One can obtain a very good estimate of discharge knowing nothing more than the general character of the plant. Let's take a pulp mill. Given the process that the pulp mill uses, one will know how many pounds of BOD will be discharged per ton of pulp produced.

There is a need for independent checks. I think these checks are partly the responsibility of government. Government, in my view, at the local level other than at the federal level. And there is also the opportunity for citizens' groups to make these independent checks. They have, in the past, provided a very valuable service in this way. I guess the other point is that one can now, or the technology is evolving, to use remote sensing from various space vehicles, either aircraft or satellites, to achieve much better monitoring capabilities than we have today.

Five years from now, I don't think it is going to be necessary to have the gauge or the instruments at the pipe coming out of a plant to get a good idea of what comes out of that plant. The combined information obtained from remote sensing, plus a knowledge of the plant process, will give the kind of data that would be required to institute a user charge system of a sort that I discussed.

DR. MRAK: Dr. MacDonald, you may recall, in California, we have a Department of Weights and Measures. They don't go around everyday, but they check on them periodically, and I could visualize a proposal you made being handled by a relatively few people going up and down checking these things.

DR. MacDONALD: I think this is an extremely important point that many who have argued against the institution of economic system, a user charge system, have claimed that it would require a vast bureaucracy. I claim just the opposite. It takes far fewer people to conduct the kinds of checks that Mr. Clark has spoken of than to administer the fifty-thousand plus permits that will be involved in the national elimination of discharges or permit system or whatever EPA calls it these days.

MR. CLARK: Do you know any place in the world, Dr. MacDonald, where this is being tried out?

DR. MacDONALD: There has been, of course, some long-term experience in the Danube region, not precisely along the lines I would argue for, but there is some experience there.

There is developing experience in Czechoslovakia and East Germany in connection with water management on the Danube and its tributaries. I think these are international experiences that we would do well to study far more thoroughly than we have in the past.

DR. SMEETS: I was very impressed by the many pertinent statements of Dr. MacDonald, and since English is not my native language, I might have some misunderstanding of what he said.

What I do not want to do is defend EPA's situation in this respect, but what I rather prefer, since I am here and I am the official of the Common Market countries, is to wonder what is the similarity between the problems of EPA and us in the Common Market.

Dr. MacDonald has referred sometimes that since the regulatory requirements you have might be too severe that people might go without work because the industries will have to be closed. But now, I try to find out the similarity between EPA and the Commission of the European Communities (the Common Market).

Is there not any problem here in the United States that is similar to what we call the technical trade barriers? I could imagine,for example, that when there are no common rules for the federal states, the United States, or in our case, for the Common Market, that one state, in the United States, or one country in our situation, that has other rules than the other one, that there could be technical trade barriers. I mean that one country, one state here, could produce cheaper than another; and that for this reason, you might close down industries since the production is cheaper. So, I think that is the situation here, and that EPA is making regulations without consideration of the individual states' regulations. It must be doing this. I am not a bureaucrat, I am rather liberal, too, but there are situations where you have to do something to harmonize the development between the different states in this case here or between the countries in our case. Therefore, I do not agree with your pertinent statement to liberalize all this development. I think you have to do something.

What about interstate rivers as in our countries, the intercountry rivers. You were referring to the Danube, but I might refer to the River Rhine. And, as I said previously in other meetings, I do not want to stick my neck out here, because of all the troubles we have with the River Rhine. But, I must tell you that harmonization between the countries is absolutely necessary. I should like to have you comment to this particular field. So to your particular statement, I cannot agree, I am very sorry. I think Dr. Talley will not do it either.

DR. MacDONALD: I think that the problem that you have referred to is a crucial one. It is one, again, one when I was in government that we devoted a great deal of attention to, and actually, through the OECD, did achieve a statement of OECD countries that environmental regulations would not be abused to develop non-tariff trade barriers. That, in fact, the principle that I mentioned in my presentation, the polluter should pay, was adopted in principle by the OECD countries. That, to me, implies that government will not subsidize industry in order to achieve pollution control.

There are many ways in which one can work at the polluter and who should pay the principle. One is through regulations, as we have proceeded in this country, in which the implementation of the regulations is the responsibility of the regulated industry. Another way the polluter should pay; he pays if he pollutes, that is, he is charged for discharging. And this is an alternative approach. I quite agree with you. I think there is every need to harmonize environmental means between countries. Whether those means are regulations, as they are for the most part today, or where they are users charges, as I hope they will be in the future.

But harmonization doesn't mean that every country must adopt identical methods or ways of approaching the problem. Within the common market, you face very special problems, because of the nearness of the countries, the number of countries, the very special trade relations that have developed between those countries. But at the same time, I think that the common market and the United States of America and other members of the major western trade nations must work towards a harmonization of environmental controls, whether they be regulatory or economic.

DR. SMEETS: You mean only on the level of recommendations?

DR. MacDONALD: On the level of recommendations.

DR. SMEETS: It doesn't work.

DR. MacDONALD: I argue that hopefully it can work, if, it has worked to a limited extent with the experience through OECD. Case in point is harmonization of regulations with the use of polychlorinated biphenyls, a particular chemical that has possible harmful effects. OECD countries have agreed to limit the open use of that particular chemical. I think that that kind of approach is the way we should go, rather than attempting to achieve international regulation, much less national regulation.

DR. SOMMERS: I was very interested, of course, as a government official, a federal government official from Canada, to hear Dr. MacDonald's views on laissez-faire economy and how effectively they work.

But you see, Dr. MacDonald, the examples you gave were from Czechoslovakia and East Germany. Countries which, in fact, hardly have the uncontrolled economies that you were suggesting would best serve the control of these industries. I think that the basic problem with these polluters paying user charges is that essentially they've got to be regulated. And, they've got to be regulated in the same way as any other system. Now, it may be there will be fewer regulations or fewer laws or fewer permits, but there still has to be regulation. Now, we then ask, who regulates them. I think that you are suggesting that the polluter industry regulates them.

Now, the sad and unhappy experience through western society for over a hundred years is that if we want to have health centers, we simply cannot rely on the polluter or the particular industry to provide that control. We cannot rely on them to provide that control. I can think of all the examples we have locally in Canada in terms of occupational health. We have to provide control.

The other suggestion you have is local control and, there again, I am less impressed. I have seen local control in terms of some of our recent problems of mercury pollution and chloralkalide plants, for example. Many local officials will tell you, they won't say it too publicly, that they are more susceptible to economic pressures than those of us one level away at the federal level.

And, though we may not do as effective a job federally as you would like, I am not very certain that the other alternatives that you have offered, in fact, will provide any control whatever.

DR. MacDONALD: I certainly agree with you that there are difficulties that imposition of user charges is really a form of regulation. It is a different form. My defense of it is on grounds of efficiency. Economically, I think it would turn out to be more efficient than attempting to regulate by the method this country has adopted. I am familiar with, I think, the progress that you have made in your country, and I see very great advantages to the particular systems that you have used. I object very strenuously to the way we have gone at the permit system within our country. I think it is going to turn out to be a highly inefficient way that is subject to all kinds of pressures of the sort that I have mentioned.

DR. TALLEY: EPA must have a word. I will only use examples that have already been mentioned. It is interesting that the ICC (Interstate Commerce Commission) was used as an example of inefficiency, but, if you will take a look at the way the ICC really operates, it is the regulated method being disputed. For instance, if you have ever had to move your household goods in this country, you discover that essentially, the federal government acts as the negotiator, so that the price is fixed among all possible movers, you begin to question the idea that the regulated people can set their own regulations.

PCB's were mentioned as examples of splendid cooperation. I think Dr. Sommer would agree that with the U.S.A.-Canadian border and the ban on open uses, we have achieved great results, but the recycling of the PCB's and then shipment over the borders from Canada to the U.S.A. and vice versa, still leaves us with a problem of PCB concentrations in the fish in the Great Lakes.

Dr. Junk asked Dr. Coulston and mentioned the fact that you cannot produce dose response functions. Sticking with PCB's, I think we can get the risk. The mink producers in the Great Lakes regions have known that for some time, if you feed the fish from the Great Lakes to their mink, they won't breed.

It is quite easy to make economic calculations of the effect - the risks of using certain contaminants. What people fail to recognize, and now I come down on the strict environmentalists, is that at some level you may decide that the risk is negligible. Especially after a great deal of dilution in your water supplies.

The one thing that you cannot do is continue to use the concept of the threshhold. We have got to forget that, in terms of public health. We are going to have to go to blind arithmetic and linearity and all sorts of assumptions, as has been done with the radiation limits that have been mentioned. And, here you know that we are going on supposition and guessing. But, when we do that, I think we should be willing to admit that that is what we are doing. That there is little scientific validity beyond an urge to protect the public health and welfare.

DR. EGAN: My comment is a very brief one, and it really relates to Dr. Sommer's criticism earlier of the effectiveness of voluntary control of pollution by industry, from which he said the federal governments have to provide controls. Well, I might differ with him very slightly in saying that they have to be prepared to provide controls.

DR. BLAIR: I believe that we should give serious thought to what Dr. MacDonald has spoken about, especially for the free enterprise system. Industry is highly responsive to economic forces, and so are the public, in general, concerning jobs or prices or whatever it may be. The large segment of the R&D potential of the nation resides within the industrial confines, and, as such, when proper economics can be brought to bear upon the solution to a problem, that is where they will tend to go.

In contrast to a strictly highly regulatory system in which you then must employ the skills of the legal people, there is not a whole lot that R&D can do. So, I for one, believe that we should give some consideration to what Dr. MacDonald has spoken about. For industry itself is not really interested in developing a highly polluted environment that it, in itself, must live in. So, I think we are all working towards the same goals, and it is a matter of a proper balance of all these ways of getting at it.

DR. COULSTON: I think it is worth giving a few examples. In the food industry, the Food and Drug Administration of our country allows the industry to police themselves. They watch over it, but it is impossible for the Food and Drug Administration to go into every food factory, every drug factory , in the United States. They would have to have hundreds of additional people, and that is no exaggeration.

But, they depend on the industry as a group, not as individual companies, to police itself. The perfect example is the Flavor and Extractive Manufacturers Association. They police themselves so carefully that if anybody sells a vanilla extract that contains vanillin, a chemical synthetic, they will, themselves,,fine the company. They do not wait for the Food and Drug to tell them that it has been adulterated. They police themselves.

I don't know why this kind of a concept couldn't be brought together. Business and government working together for the benefit of both and the consumer. That, I think, is the necessary ingredient.

DR. BUCKLEY: I wanted to take exception with a number of things that Dr. MacDonald had to say. The biggest exception I would like to take is, I think it is a gross over-simplification. And because it is, and speaking for affluent charges or economic mechanisms for control, it seems to me that it effectively becomes unworkable. All the objections that we have heard, in a way, speak to this simplification of regulation on the one hand, or economic charges on the other. Now, I would have responded in part, that the reason that we have the laws that we do, is that the only thing legislators can do is legislate. And, in fact, that is what they have done. We have innumerable laws in this country and will probably have more.

The point I wanted to make, in part, is that it seems to me that some sort of a hybrid system between regulation of those things which people cannot see, or detect, for themselves, probably need to be regulated in a quite different way than some of those others where the polluting effect, if you will, is obvious. If the stream stinks, if it is dark brown, if we have those "lovely" bubbles that Dr. Coulston spoke of, it is obvious to the citizen that this is the case. And then, they are in a position to take some action to exert an individual choice.

On the other hand, if it is a substance which is present in fractions of a part per billion, you can't smell it, or taste it, or even by most chemical means, detect it, clearly, the individual has no freedom of choice as to whether he will take this in or not, on the basis of his own experience. So, I plead with you, all of you, as you go through this meeting, to phrase your thoughts sufficiently well that the over simplifications don't seem to be what we are really driving at. Because it is clear to me that it is not an either/or proposition on how we will best go about this.

MR. ANGELOTTI: I am with the Food and Drug Administration. I would like to add a comment to that, if I may, and respond to Dr. Coulston's remarks of how Food and Drug attempts to conduct it's business.

I think it is important that you recognize that agencies such as ours and EPA will more and more find itself in a position similar to FDA's, as time goes on. That agencies, such as ours, have to get on with their jobs the best way that they can. There is no either/or situation, as was indicated. You use whatever tools are available to you to get the job done.

In some instances, and these are few, one can depend on industries to self-police themselves. However, even in such a situation as that, there is a bureaucratic need for monitoring the effectiveness of that self-imposed industry system.

In my view, the control that the Food and Drug Administration has exercised over the food industries in this country is a result of very strong regulatory activity that has occurred in the past. As a result of that strong position adopted by the Food and Drug in the past, it is now possible to convince industries that self-policing may be to their benefit. But, in the absence of a strong regulatory posture, in the beginning, it is very doubtful that industries would adopt a voluntary system. It has been our experience that they will adopt such a system when there is an economic advantage to do so. Such an economic advantage is brought to their attention through a very strong posture on the part of the regulstory agency.

So, I would say, that there is a mixture of approaches that are possible here, but to go forward initially with a concept of voluntary controls, I think, would not effectuate the congressional mandate that was laid down upon the EPA, and which is, theoretically at least, the will of the people of this country.

DR. MIETTINEN: There are facts, or opinions, which Professor Korte presented that showed a tabular increase of the trace levels of various contaminants on a different production increase basis. For example, he stated that the possible increase for ten years of a two percent per annum production assumed no changes. Can I ask Professor Korte whether that included no degradation, no chemical changes at all, or if it did, what the situation might be guessed to be if it did include chemical degradation?

DR. KORTE: To answer this question, I would like to explain the assumption. These figures go back to 1875 and assumed that at that time that there would be no, for instance, PCB present in the environment. Now, we find that the global concentrations of PCB are very well distributed.

If this continues to go on for the next hundred years, including all degradation which might have occurred during the last hundred years with the existing PCB, then you might expect, if you have a production increase of two percent, a concentration a hundred years from now in the order of the same degradation level as we had during the last hundred years. I don't know whether it is a reasonable assumption, but it is the only thing we could do.

If it comes to a higher concentration, then possibly the degradation has a higher rate or lower rate. Nobody can predict it at the present time. But the data shows, with reasonable assumption, we are in the face of global pollution, and have a hundred years' time to solve the problem with the so-called chlorinated hydrocarbon chemicals.

DR. SOMMER: Dr. Korte, you presented a table showing the contaminants among other things by growth to 1974. The highest concentration of pollutant was, I think, at twenty one hundred parts per trillion of tetrachloride ethylene; I think I remember that from the table. The question I want to ask you, is an industrial effluent, or was that formed by chlorination of the water? Do you have any evidence as to how it is present?

DR. KORTE: I would think this is production directly. It comes out from the factory, not from chlorination of some ingredients in water. But, just from the production level. It is my assumption, and I couldn't even think how it would be formed in water at this level of normal chlorination.

DR. COULSTON: Dr. Korte, you discussed the point of view that one part per million was perhaps the level at which biological effect would become significant, or much higher than that, and therefore, you would not care to pay attention to anything below one part per million.

My own concern is that indeed, there may be effects at this lower level. Are you saying there are no effects? Do you feel that it is legitimate to ignore them at less than one part per million? Do you have any sense of a factor of safety that might be appropriate for certain of these chemicals?

DR. KORTE: I like to always take a pragmatic point of view. We know only the global occurrence of organic chemicals in this world, since no longer than eight years. That means that we learned it from a few examples, for instance, organic chlorines. These few organic chlorines present less than one percent of the total chemical production. That means the greater part of chemicals are still in the dark. From these chemicals, from these organic chlorines, we have the knowledge that concentration of PPB area do not show any reasonable biological effect. At least, that is my knowledge.

We have only two examples in humans. I am speaking of humans, if I may. We only have two examples of diseases which occur with humans following chronic intake. And these are two examples from Japan, which are known for the past eight or ten years, caused by mercury and cadmium.

And, in these two cases, the intakes were in the order of ppm. To my knowledge, there is no other disease of this kind to humans shown with any concentration below ppm. I do not even know of any disease, chronic disease, environmental disease, of this kind occurring below the ppm level.

But, generally speaking, if you consider all the chemicals which were released over a hundred years into the environment, we know only a very smal part of the total. In spite of that, we only have found these two chronic environmental diseases, on this basis, I would think we have more things to do and better things to do than to worry about concentrations of a ppb or a ppt. I think the bigger problem is to get all the big products under control, which we release day by day into the environment, which we just neglect these days.

DR. KIMBALL: Dr. Korte, again. Your comment relative to waiting until one observes human disease as a result of environmental contaminates is, in my view, a shocking statement. What does one do with the animal data that are before him, indicating that certain chemicals, at least, in the ppb range, do have affects on animals? Until we have better animal to man bridges, what are we to do with such information? I don't think one can afford to ignore such information. I don't know how you deal with it exactly, but you can't ignore it. And, I gather from what you are saying, that you would lead us to ignore such information.

DR. KORTE: It was not my attempt to ignore anything. But, you know, two weeks ago, we had a meeting in Munich, discussing interpretation of animal data to man. It was a big, complicated problem, and not well understood, scientifically.

My only point, which I tried to make, is we should not concentrate our whole scientific effort worldwide, just measuring PCB or DDT or whatever is in ppt concentrations. We have greater objectives, greater tasks before us, because these compounds only represent less than one percent of the total chemicals released into the environment. It is hard for me to understand that everybody in science, or out, is talking about PCB and DDT. If you ask for research programs, you go around to the universities or government research institutes, that everybody talks on mercury, on cadmium, on possibly selenium, or something else. Let's name just five metals, that's all, the others don't exist. And, of the organic chemicals, everybody is talking about DDT, PCB, hexachlorobenzene, and chlorinated hydrocarbon pesticides. These chemicals can be measured very easily, and everybody proposes research along those lines to magnify the sensitivity, or who knows what.

I think we should not neglect to get the whole problem of the release of organic chemicals under control. That is my only point. And, we should save a little scientific effort for this direction. I should not talk so much about toxicology, I think there are more knowledgeable people here who could answer you about the toxicologic point of your question. But, actually, while there are only two environmental diseases, it still sticks, stays. Two environmental diseases with chronic intake know, which are dangerous to man, to humans. These are very local diseases, not global diseases.

DR. SMEETS. I have been listening with great interest to the discussion taking place. Recently, we had a scientific meeting on the relationship between the hardness of drinking water and public health.

Listening to Dr. Korte, I still have to question what is natural water, that is my first question. The second question is, is it useful, and is it wise, that we globalize the quality of natural water.

When we compare geographically the hardness of drinking water and the inverse relationship with cardiovascular diseases, so that you can draw a line from, let's say, North England to Southeast England on one hand, there was an increase in mortality of heart disease in the Northwestern part. We also learned from WHO, during the same meeting, that a similar line can be drawn from the Northwestern part of Europe to the Southeastern part, and an increase in cardiovascular disease was found in the Northwestern part, in comparison with the Southeastern part of Europe. Therefore, we see that there is a great difference in adaptability of the population to the different types of water, even so-called natural water.

So, my question is: first, what is the definition of natural water; and second, should we generalize this for all the world in comparison to the adaptability of the people?

DR. KORTE: To my understanding, there is not any definition for natural water. Natural water is just the water which comes out of the sources, and this can contain many chemicals. It means we have to realize that natural water is the water that comes out of the earth, or falls down as rain, that's all.

If you find some correlation between some constituent of this natural water and some disease, then take out the constituent. It is just a question of investing money. Technologically, it can be done. And, it can be done on a regional or city-wide basis, or whatever way you would like to do it. Again, the pragmatic approach.

DR. TRUHAUT: Ladies and gentlemen, I have to refrain from making comments or asking questions. As Chairman of this morning's session, I have to give the floor to other people. But it is my duty, as Chairman, to conclude this session at this time. The papers were excellent, the discussion was warm and, for this reason, I congratulate again those responsible for the program and the Forum. I think I have to send you to lunch, to take some caloric stimulation. I wish you good appetite, and thank you all, the speakers, and all other participants.

Lunch - Speaker: Dr. H.G.S. van Raalte

What qualifications I may have for talking to you in a water quality program, I share with practically all people of the world. For 70%, I am made of water, and a pretty good quality of water, at that. That fascinates me, because that is the water which ultimately determines how we feel, how healthy we are. Yes, as a private citizen, I know there is more about the quality of water, as I said earlier, but as a physician I am, as I must be, concerned with human health. 70% of our body weight is water, 50% in the cells, 20% extracellular, outside the cells. Everyday, apart from what we drink, our daily intake of water, with or in our food, is 475-725 gm. A further contribution to the daily intake is 250 gm of water of oxidation. On the other hand, our daily waterloss is 125 gm in stools, and 625-850 gm by evaporation through skin and lungs. With drinking one liter per day, our daily intake of water is about 2 L. That means, an intake of water by humanity world-wide of two times 3.6 10^9 liters. That is 7 million tons of water daily, or seven of the biggest mammoth tankers full.

Now, if this water contained one hundredth of a ppm of a pollutant, that would mean one hundred tons of that pollutant for humanity everyday. These figures themselves are certainly frightening, and may well serve to scare the public. It is, however, not the way these matters ought to be considered.

I am not a scientist, whatever that may be. One of my Swiss colleagues once said: "A scientist is not quite an exceptional person. He does not know more than most other people, only he has his thoughts well organized and uses slides." Not today, but tomorrow, I shall be using slides. That does not make me a scientist, however.

I am only an old country doctor. I have really carried out the practice of medicine for half a lifetime. I have treated,such as it was, the sick and imaginary ill. I have had blood over my hands and tears on my shoulder. Among the many things I learned, and the few things I now know, is when a person is ill and when he is not. Much of this I learned from a headsister in one of the hospitals where I have had the privilege of practicing. I remember many patients in whom laboratory test results represented just so many bad omens, pointing to an immutable fatal outcome before the next sunrise. I remember on many such occasions this headsister telling me: "Come on, Doctor, you are tired. You go on home to bed. This man, or woman, will be alright. Don't worry, I'll call you if anything goes wrong, you know that, but he'll be better tomorrow morning." She was almost always right, and I was almost always wrong.

And the reverse was also true. After my late rounds, looking at the day's laboratory reports, I said: "Thank heavens, this patient looks much better. He is going to make it." And then she might say, "Doctor, you had better come back before you are going to bed. This man has me worried. I don't trust it." She was almost always right, and I was almost always wrong. Until I learned that, to not always trust the laboratory alone. Now, please, don't misunderstand me. I do believe that laboratories are extremely important, in fact, they are essential to everything concerned with human health. However, just like my friend, Johannes Clemmesen, from Copenhagen, used to say when speaking about statisticians, so it is with laboratories, "They are good servants, but bad masters."

There is one other thing I would like to say. Nothing in life is black or white. There are no things like that, no matter what, are only bad for you, and there are no things that are only good for you. That includes oxygen, which is pretty toxic in high concentrations, and salt, and water, and even food and love. However, the latter two, you've got to choose pretty carefully. Many years ago, I was in Iran with a survey and pioneer party, in which I was charged with looking at requirements for medical services, facilities and supplies for the consortium of the eight big oil companies. This was a small party, and some of us lived in a guesthouse, where we had our meals together. The work, at times, was a bit frustrating, so some of us ate a bit more and became a bit more obese than would be desirable. Inevitably, table talk turned to slimming and dieting, on which I, as being the only doctor, was considered to be the expert. A nicely-rounded little girl among us, a confidential secretary, one day asked me what she should do and eat in order to lose a little weight. So I told her, only half jokingly, "You'd better live on love and peaches."

Now those days, I had to travel quite a bit, all over the country, to the refinery at Abadan, to the various oil fields, pipeline pumping stations, and harbours, and it was a good four weeks before I returned to Tehran. Those days, air traffic was not as well organized and reliable as it is today, detours and delays were not unusual. When I arrived at the guesthouse, it was well past midnight, there was a power failure, and it was very dark. When I opened the door, there was the little love-and-peaches secretary, putting her arms around my neck and confided very excitedly, "Doc, it worked! I'm engaged!"

When I was saying that there are no things which are only

or always bad for you, I was referring to the dose. A contemporary of Columbus, Theophrastus von Hohenheim (1493-1551), the famous Paracelsus already wrote, "Dosis sola facet venenum," only the dose makes the poison. If the dose is large enough, everything is poison. If the dose is small enough, nothing is poison.

This is as true today as it was in Columbian times. And it applies specifically to the water. In modern times, clearly the ability of the analytical chemist to find the presence of a poison in the water has outstripped the ability of the physician to find a disease to fit the dose of the poison. Every poison, ever discovered, can now be found, and is being found in the water, but apart from a few infamous examples like Minimata, where concentrations were considerable, there is no evidence that the water-using public is any less healthy than they were before the chemical era.

Neither is there any evidence for the mythical assumption of the existence of highly susceptible groups in the population. I know of no drug or chemical that causes disease in babies, women or the infirm in a dose less than one-tenth of the dose that is a no-effect dose in the healthy.

Therefore, I would like to say, as Dr. Talley said, "Let's keep our priorities right." Let's not spend our limited talents and resources to fight the windmills of those residues in the water that are toxicologically insignificant. But rather, let us use our efforts in those areas in medicine and public health where we can relieve real suffering. But that would be another story.

When asked about improving the quality of water, though, I believe the best advice ever given comes from this country. In a simple advertisement: "Nothing improves the quality (or was it the taste?) of water like our 'So-and-So' whiskey." I am not allowed to mention brand names, of course.

I have not been talking as a representative of anything or anybody, not even of the Windmill People. Consider what I have said as just some thoughts of an old country doctor.

OPENING REMARKS - AFTERNOON SESSION

DR. EGAN: Can we start the afternoon session, now, please? Thank you, Monsieur President. May I, in opening the afternoon session, say a word of thanks to our President, Professor Truhaut, the Chairman of the past morning's session. I am not quite sure how many of you appreciate that he only arrived in Washington from Paris late last night, and that at the conclusion of these meetings on Wednesday, will go straight back to Paris, in order, then, for the two remaining days of this week, to take the chair of the Scientific Committee of the European Economic Community. You certainly stimulated us, and activated us this morning, Mr. President. Thank you, very much.

It is difficult, as already said by Dr. VanRaalte, for anyone to follow Professor Truhaut, and I would like perhaps, briefly, for those of you who were not able to be here first thing this morning, to say one or two simple mechanical things, most of which, I think are now evident, but not all of them. That everybody is free to participate in the discussions, whether you are a speaker on the program with a name in front of you, or whether you are an observer here, you are free to join in the discussions, which are being tape recorded. You have the opportunity, if you wish, to say that you wish to be off the record. So, please feel free to join the discussions.

This afternoon, we begin to look at more specific areas of interest in relation to water quality. We begin, in fact, with something still fairly general, parameters for water assessment, in a paper by Professor Doctor Sontheimer, from the University of Karlsruhe, in the Federal Republic of Germany. The paper is being presented by his co-worker, Dr. Wolfgang Kühn. Now, it is quite true that Dr. Kühn spent two months (ten weeks) here in the United States, earlier this year, on the West Coast with EPA, and he is in the program attributed to the EPA, but I think that is the only claim he has to being with the EPA. He is a co-worker of Professor Sontheimer, and I invite Dr. Kühn to present to us now, "Parameters for Water Assessment".

PARAMETERS FOR WATER ASSESSMENT

Heinrich Sontheimer and Wolfgang Kühn

Summary: Defined are the I.A.W.R. Standards of the Categories A and B for the River Rhine. The maximum and average values measured at the lower Rhine in 1974 are given. Examples prove, that it makes sense to establish sum parameters for dissolved organic material as standards. The behavior of the sum parameters with the flow indicates much about the kind of behavior during treatment. Data for group parameters, as organic chlorocompounds, organic sulfur and humic acids, are given.

In the I.A.W.R. (International Commission of the Waterworks on the River Rhine), waterworks from Switzerland, France, the Federal Republic of Germany and the Netherlands are cooperating. Some years ago, this organization worked out water standards for the River Rhine under the conduction of the Engler-Bunte-Institut Department of Waterchemistry at the University of Karlsruhe. A distinction was made between stream standards of categories A and B which are defined as follows:

A Standards: These should be observed if only natural water treatment facilities are available; that means simple biological and mechanical processes. Up to twenty years ago, sand bank filtration was the only treatment used along the Rhine for a good drinking water quality. Therefore, the A parameters should be a goal for wastewater treatment plants discharging to a surface water which is used as a drinking water supply.

B Standards: In this case it is presupposed that the wellknown physico-chemical water treatment facilities are applied in addition to river sand bank filtration. This means treatment with ozone, floculation, filtration, granular activated carbon adsorption and disinfection.

Table 1 shows the comparison of the I.A.W.R. Standards with the maximum and average values, measured at the lower Rhine in 1974. The parameters are grouped according to general measurements, like oxygen demand, conductivity, color, taste and threshhold odor number, suspended organics and inorganic water contaminants - not further mentioned in this context - and organic contaminants, for which a number of sum and group parameters are prescribed.

This paper deals with the importance of these parameters with respect to the raw water quality. The reason for using only sum and group parameters as standards instead of standards for single organic substances will be explained. Measurements of the concentration of single substances are also taken as shown in Figure 1.

Figure 1 shows a gas chromatogram of an activated carbon extract. The sample was taken out of a Rhine water work-filter, and it shows a high number of hygienically and toxicologically objectionable organic chloro compounds. Figure 2 shows a relatively big variation of the concentration of a single organic compound, for example, here Dichlorobenzene.

Frequently, the observed maximas are ten times higher than the average concentration. Peaks in this range complicate the control of the desired standards. The standards are usually considered to be applicable to the maximum and not to the average values which occur in a time period. In addition, the sum of the individual compounds which have been determined reaches barely more than 0.1 to 1 percent of the total organics. Also, the analysis of single organic substances is complicated and expensive and therefore not practicable for the daily routine control in a waterworks. The practical experience at the River Rhine has proved that there is no direct correlation between the discharge of any single compound and the difficulties occuring during drinking water treatment. But there is a relationship between the total organic carbon and such problems, or between some group parameters and these problems.

The specific conditions at the River Rhine will now be examined. Five years ago, more than fifty waterworks initiated a research program investigating the pollution of the Rhine in respect to the drinking water supply. A summary of these results is given in Figure 3.

Here, the geometrically averaged data for the flow rate, the DOC (Desolved Organic Carbon), the COD (Chemical Oxygen Demand) and the UV-Extinction are presented. The pollutant concentration increases from the upper to lower Rhine. In the past four years, the high of the sum parameters has been decreasing continuously. However, the flow rate has been increasing in this period too, and therefore the apparent decrease could possibly be explained by dilution. In order to eliminate this influence of flow rate, the pollutant loads are calculated from concentration and flow rate and compared with each other. But this method is only reliable if the total discharge of the dissolved organics is independent of the flow rate. This is not right in most of the cases, as shown in the Figure 4.

Here, the average values are sketched for the DOC-load of the Rhine at Benrath for the past five years a function of the flow rate. The load is increasing with flow rate. That is the situation all along the Rhine, even at the relatively clean upper Rhine at Basel as shown in Figure 5. The curve is calculated by the method of the minimum squared error for the years 1970-1973; in addition, the data for 1974 are presented. In contrast to the lower Rhine, shown in Figure 4, it is apparent that in Basel, the quality was better in 1974.

The equalizing curve used for these data is an easy-to-handle method for determining variations of the amount of pollutant as a function of a river's flow rate. An example of the variation of the DOC-load at the same location of the Rhine at Basel is given in Figure 6.

The curves for the years 70-71 and 70-72 are compared with those for the period 70-73 as well as those for 70-74. Typical systematical differences are obvious, leading to a shift of the curve in certain regions. In this case, a decrease of the DOC-load and therefore of the DOC-concentration at the same flow rate was observed.

In Figure 7, a comparison is made of the DOC-load and COD-load for the different parts of the Rhine. The percentage variation of the average DOC and COD values along the Rhine is compared with the calculated content in the years 70-72. In every case, there is a small improvement of the COD data, probably mostly caused by the increased flow and the accompanying changes in humic acid concentration of the natural water. The changes in the DOC are different in several parts of the river. In the upper Rhine there is an improvement, while at the middle Rhine the results worsen. However, the changes are not large, but the measurement of DOC is very exact and is an excellent indicator.

There are two ways to judge the changes in these organic sum parameters. Either more information can be obtained from the data by mathematical calculations or more parameters can be measured using newly developed analytical methods.

First, one can try to create a model which explains more than the measured data. In this context, we were trying to explain why there is such a change of the DOC with the flow. With increasing flow rate, the flow time of the water decreases. The flow time between two points on the Rhine is inversely proportional to the flow rate to the power 3/7, as the measured data indicate. Plotting the values of the DOC-load against the flow time determined, this manner yields with good approximation a line which allows further interpretation.

In evaluating the data, it can be presupposed that the amount of material which can be biologically degraded is proportional to the concentration of biodegradable substances as well as to the flow time. The method provides a correlation between the total organic load and the load of degradable material. In this way, you also get a number for the bio-resistant organic materials which are especially important with respect to drinking water treatment. The next diagram shows that the function developed by this method describes the measured data for the DOC in a river very well (Figure 9).

All the DOC-loads from the lower Rhine at Wesel near the Netherlands are shown. The proposed and sketched function reveals a good agreement with the measured values.

A computer program using as input the measured sum parameters gives easily the different constants for the equilibrium curve, which leads to a value for the load of bioresistant material. The change of this parameter according to the flow distance is documented in the next picture (Figure 10).

The load of bioresistant or slowly degradable organic material increases downstream. Comparing these values with those for the total DOC impact at different flow rates leads to the conclusion that a remarkable part of the total amount of DOC belongs to the category of bioresistant substances. Some of these are only partially degraded during sand bank filtration. For this reason, they are a very important measure of the raw water quality with respect to later use as a drinking water resource. Beside this, some of the compounds belonging to this bioresistant group are toxic and hygienically objectionable. As shown, this important parameter is evaluated by measured organic sum parameters by the new equation. For the calculation, the DOC results are very suitable because other parameters vary too much, e.g., the COD. Other parameters, like BOD (Biological Oxygen Demand), are not useful, because the BOD measurement at the Rhine includes a large nitrification oxygen demand because of the high ammonium concentration.

The conclusion of the evaluation of the sum parameter DOC is shown in Table 2. You see the average value of the DOC and the data for the resistant organics (DOC_R), calculated by the described mathematical method. At the Rhine, the content of these compounds, based on the yearly average, is between 25 and 55 percent. Further research work proved that values around 25% are mainly caused by the presence of humic materials which are natural bioresistant organics. Higher values from the Danube and the Rhine at Karlsruhe are due to papermill wastes and the hereby introduced bioresistant ligninderivatives, often chlorinated lignites, caused by bleaching. These substances make drinking water treatment more difficult.

These examples prove that it makes sense to establish sum parameters for dissolved organic material as standards, especially the DOC. The behavior of the sum parameters with the flow indicates much about the kind of the organic compounds which are present and their behavior during treatment.

We have worked on the development of new analytical methods for describing the most important groups of nondegradable organic compounds and for measuring their concentration. This has been accomplished for several groups. First investigations with the new methods were made for the year 1974. Without explaining details of the methods, results are listed in Table 3.

For three locations on the River Rhine, the DOC and the activated carbon adsorbable organics are shown. The adsorbable part is above fifty percent and is increasing downstream. Organic acids are subdivided into ligninsulfonic acids, other sulfonic acids and humic acids. The ligninsulfonic acids, as a waste from papermills, are important and reach 22% at Cologne. The humics are on the average a third of the total DOC.

Of hygienic importance is the content of organic chloro compounds. Most of them are non-degradable and the concentration increases as the length of the river increases. A lot of them are generated during the bleaching process at papermills. Another very volatile part stems from dry-cleaning. Organic chloro compounds are used as softeners or for agricultural purpose as pesticides farther on. A lot of them are only intermediate industrial products. They are also produced during high prechlorination in some waterworks and in waste water treatment plants. None of them are from natural sources and therefore they are an indicator of industrial pollution. Especially the aromatic nonpolar chlorinated compounds are stored in the body fat and are highly toxic and carcinogenic. The concentration at Basel at the upper Rhine is already very high and remains almost the same throughout the river. That means there is around one percent, and relatively more after sand bank filtration because they are not degraded or removed during this step.

A change in the flow rate is of less influence for the load of organic chloro compounds, sulfonic acids and ligninsulfonic acids, but the load of humics and of biodegradable substances is dependent on the flow. Using this as a basis for the calculation of concentrations at other flow rates leads to the information shown in Figure 11.

Figure 11 shows the dependence of the main group parameters of the dissolved organics on the flow rate. It should be observed that the total amount of the DOC is going to be reduced by a factor of two if the flow rate is increasing from 800 to 4000 m^3/second, and that the fraction of the non-degradable compounds is also becoming smaller. This small concentration of DOC at high flow rates, and the fact that in this case humics and degradable substances are the main components, results in the fact that there is no problem with dissolved organics in water treatment plants under these circumstances. The load of trouble-making substances is so small that the treatment that is used is successful.

The situation worsens at low water levels, however. First, there is a high total DOC and second, the part of industrial, hard-to-remove contaminants goes up to 60%, and this especially causes a lot of problems in treatment plants.

Figure 12 shows the problems and the effectiveness of the removal of these compounds during treatment. Given here are the values for DOC, adsorbable substances, organic sulfur and organic chlorine from the raw water, after sand bank filtration and for tap water. Especially the chloro compounds are meaningful because there is no reduction of them during sand bank filtration.

The organic chloro compounds, therefore, are a main load of granular activated carbon filters, as shown in Figure 13. The parts of sulfur and chloro compounds of the total load are listed for granular activated carbon filters located at three Rhine waterworks. Samples are taken from the top and the bottom of the filters. The total percentage removal is not very good. On the other hand, especially the organic chloro compounds are in competition with other highly toxic material for adsorption sites. With respect to this, prechlorination is critical, as Figure 13 shows in the third case.

In our opinion, based on the experience with the Rhine, the concentration limits and standards, given for sum or group parameters are significant. There is much more information in these parameters, as shown with the computer program. The reason for the limitation is not only the toxicity, but especially the fact that a high concentration of these parameters reduces the security capacity of the treatment plant. This point of view and the experience of the waterworks itself, was the reason these parameters were established as standards for water quality at the River Rhine.

PROBLEMS OF BOTTLED WATER

J. T. Hutton

INTRODUCTION

My remarks this afternoon represent the views, experience and concerns of one company. The Foremost Foods Company. A company which through its wholly owned subsidiary has been in the bottled water business for more than fifty years. I will, however, make no attempt to recount the whole fifty years of our experience in the processing and distribution of drinking water.

In keeping with the theme and objectives of this forum, I will limit my remarks to three points:

First, to place the subject in perspective, I would like to sketch the *what* and *why* of bottled water as a business.

Secondly, I would like to examine certain problems facing, but not necessarily unique to, the bottled water business as a supplier of drinking water.

Thirdly, I would like to explore a few thoughts which may point us in the right direction for finding effective solutions to these problems.

The bottled water business has a simple base. The consumer. To day, as in the beginning, our products are designed, built and delivered in response to consumer wants and needs.

Consumer research has been described as a "way of life" with our bottled water operations. This close rapport with the consumer, no doubt, explains why bottled water consumption continues to grow despite the price advantage held by municipal water which is more conveniently available, either hot or cold, at the turn of a tap in nearly every home in America.

Why consumers pay a premium for bottled water is explained by the kind of products they say they want and, more emphatically, by the kinds of products they buy.

In our experience, the consumer wants one or a combination of several things which is perceived as missing or questionable in the municipal supply. The reasons most commonly given for buying and drinking bottled water are:

- Taste
- Uniformity
- Safety, and
- Health. Fluoridated water is a specific example of a water in this category. Some consumers want it, some don't. To accommodate both preferences, we offer a choice of water with or without added fluoride.

To ensure that our products possess the attributes sought by consuming public, we start by selecting the best available water supply in a particular geograph location. In some instances we use water from our own deep wells while in other locations we start with water as it comes from the municipal system. In either case, the incoming water is treated by a five-step process designed to achieve a finished product that is clean, clear, safe, odor-free, and uniform from lot to lot and day to day.

The first step is filtration, to remove all traces of suspended solids.

In a second step, virtually all minerals are removed by means of ion-exchange singly or in combination with reverse osmosis.

The fourth step takes place at the time of bottling when the product is saturated with ozone to destroy any remaining pathogenic organisms in the water itself and to sanitize the bottle and bottle cap.

Fifth, the process and product are constantly monitored by a rigorous quality control program. Recently, one of our people calculated that the product is tested in some way, at some point, in the processing and distribution system every two minutes!

A further consideration which serves, I believe, to place the bottled water business in perspective is the fact that its operations and products are now covered by FDA Good Manufacturing Practices and a FDA Standard of Identity.

This regulatory coverage is in large measure the result of close cooperation between the Foremost Foods Company, the American Bottled Water Association and the Food and Drug Administration. I am pleased to note that two gentlemen who have been active in building a positive and effective relationship between the _regulated_ and the _regulator_ are with us today. They are Dr. Robert Angelotti, FDA's Associate Commissioner for Compliance and Mr. Fred Jones, the Executive Director of ABWA.

In sum, the bottled water business has come a long way in the fifty years of our experience. It has a growing list of satisfied customers, it has a proven ability to solve the traditional problems of taste, odor, clarity, uniformity and safety relating to water quality. Its products and operations are controlled by what would appear to be reasonable, and effective regulations at the Federal level.

So what's the problem?

PROBLEM

The problem is one of concern about the future, not the past. In particular, we are concerned about the consequences of recent publicity questioning the quality of drinking water in America. More specifically, our concerns are centered in four areas:

1) We are concerned that consumers are confused and have become fearful that carcinogens, pesticides, herbicides, bac-

teria, viruses, asbestos and a host of other foreign substances are, or may on occasion, be present in their drinking water in harmful concentrations. In Dr. Talley's words of this morning, "We have alarmed rather than informed them."

How, for example, would you respond to a consumer who asks about chlorinated hydrocarbons? And whether the municipal supply contains chlorinated hydrocarbons? And whether the water from their tap is dangerous to their health?

What about chloroform and carbon tetrachloride? What about viruses and asbestos? What about the question of hard versus soft water in relation to the danger of cardiovascular problems and heart attack? What do you say when they ask if they should be using bottled water? Directly, or by implication should you say municipal water is unsafe?

However, you choose to answer, what proof would you use to support your position?

2) We are concerned that objective answers to these and related questions are, to our knowledge, not available. Incidentally, this is what we tell the increasing number of consumers who inquire as to the comparative safety of bottled and municipal water. As unsatisfying and sometimes frustrating as a "we don't know" answer is, we sincerely believe that any other response would be highly misleading to the consumer and contrary to our national interest at this point in time. And I want to emphasize that the ABWA has taken this same position.

One exception to the "we don't know" response occurs in the case of those consumers who have the scientific background to ask what is being done to get answers. In those situations we explain that we are working with the Environmental Protection Agency as well as the Food and Drug Administration in the use of the best available technology to detect and remove all harmful contaminants from our drinking water products. To this end we will take delivery in mid-October of a mass spectrograph which is to be used exclusively in our bottled water operations.

3) We are concerned that, as a nation, our research efforts appear to be unbalanced. Our observations suggest that a too heavy emphasis is being placed on the discovery and use of ever more sensitive analytical methods rather than on the public health significance of what is being measured. It would also seem that the immediate and pressing need for reliable toxicological and epidemiological studies is being by-passed in favor of new process technology to remove miniscule levels of substances, both natural and otherwise, which could eventually prove to be a benefit rather than a risk to health.

In sum, our efforts remind me of the story of a father, his son and a problem of concentration. It seems, or so the story goes, that the son came home one evening and handed his father a report card with the salutation, "Hi Dad, will you sign my report card?". The father, appalled at what he saw, followed the boy to his bedroom to explore the matter man-to-man. After some opening pleasantries, the father got to the point by asking, "Son, how do you explain your report card -- four F's and one D?" Following a long moment of silence the boy said, "Gee Dad, the only thing I can think of is that I overconcentrated on my best subject."

Somehow, I feel that we can't afford the luxury of concentrating on only our best subjects just because they are easier or even because of some short-term political appeal.

4) A final concern is the difficulty of and uncertainties which arise in dealing with two federal agencies having different but interacting areas of authority and responsibility related to water quality. And before long I suppose at least another agency, the FTC, will be telling us what we can and can't say in our efforts to communicate through the advertising media in an informative way with the consumer. I wonder if it is unreasonable to assume that it will use its own independent and differing set of criteria as it has in other sectors of the food business?

SOLUTIONS

As we turn to solutions, our last point of discussion, let me hasten to say that I have no instant answers to the problems which I have just enumerated.

By way of explanation, I believe that the issues in question are far too complex to yield to any sort of superficial or what might be called a "Band-aid" remedy.

It seems to me that solutions can come only when we are able to bring all of our resources -- scientific, legal, financial and political -- to bear in an organized fashion on an agreed upon and common national goal. If we want safe drinking water, we must first know what safe is. This is a job for the toxicologist and epidemiologist. Workable solutions and good laws to enforce them can come only after we have the knowledge to know what to do and the technical know-how to achieve a safe supply of drinking water for all Americans.

To accomplish this will, in my opinion, require a basic change of direction and a re-ordering of priorities at the National level coupled with the finest kind of cooperation and coordination among and between all agencies and institutions in both the public and private sectors. Above all, we must be willing to address and master the difficult as well as our easiest subject.

As a case in point, I do not believe that new and ever more sensitive analytical methods with or without new process technology can produce the kinds of answers we need.

As an alternative course of action, I suggest we give immediate attention to:

1) New research programs so designed and so funded as to determine the public health significance of those chemical and biological substances now known to be present in most, if not all, our water supplies.

2) Better coordination at the National level of the scientific, legal and financial disciplines needed to identify and correct problems proven to be legitimate matters of public safety.

3) Accurate and informative communications with the consumer regarding the health significance of chemicals now being reported as present in much of the nation's drinking water supply.

Obviously the job ahead will be neither simple nor easy. Sound solutions can be arrived at only after the facts of risk and cost are known and balanced by seasoned judgment. Toward these ends, I am confident that the proceedings of this Forum will be seen as a step in the right direction.

EUTROPHICATION OF INLAND WATERS

D. E. W. Ernst

The Joint Division of the International Atomic Energy Agency (IAEA) and the Food and Agriculture Organization (FAO) of the United Nations is very much concerned with the needs of developing countries. It tries to initiate the process of development particularly by the application of isotopic tracers aside with other advanced research tools. Most of this work has since been done in the field of agriculture and related topics. However, recently more and more problems of food and environmental contamination have been attacked. In June 1975, this interest turned particularly to water when the FAO/IAEA Advisory Group on Isotopic Tracer-Aided Studies of Inland Water Eutrophication and Pollution met for the first time at the IAEA Headquarters in Vienna.

One of the most prominent effects of pollution as well as eutrophication is a change in the growth of plants (macrophytes as well as algae) and in the microbial decomposition of organic material. This is the reason why most of the papers of this conference were concerned with algal or bacterial metabolism or related topics.

This paper is a survey about this meeting, particularly stressing the production aspect. An official report will also be published by the IAEA.

Let us follow the flow of energy in a lake: the bulk of the sunlight which penetrates into a water body plainly heats the lake up without major chemical change in the absorbing material. About 1% or less is converted into chemical energy (primarily hydrocarbons) by the photosynthesis of phytoplankton or macrophytes. This energy respectively the corresponding amount of organic material, is called gross primary production. About 60% of this is used up by the photosynthesizing organism (largely algae) itself to support its metabolism. The remaining 40% is the net primary production. This material represents the energetic base on which all the rest of aquatic life is flourishing. About 45% of this phytoplankton dies and is converted mainly into water, carbon dioxide, and some inorganic salts, by heterotrophic bacteria. The rest serves as food for the next trophic level, the zooplankton.

The percentages given vary from lake to lake, depend on seasonal or diurnal variations and may be subject to the changing pollution charges of the water body (e.g. Silver Springs, Florida, 1.2% of the solar energy are converted into organic material). Most lakes of the temperated zones convert less than 1%.

The primary production may be limited by different environmental factors: light (for instance, on the dark bottom of lakes), temperature (as in arctic lakes), nutriet supply, or poisonous substances. As soon as the meterologic conditions become favorably for reproduction, high nutrient levels cause mass proliferation particularly of blue-green algae, within a few days or even hours with all their undesirable effects as an increase in O_2 demand, which can easily overcharge the O_2 supply, and thus may cause anoxic conditions, fish-kills and general ecological deterioration. This is the reason why it is essential to know which nutrient, or poisonous material, is definitely limiting autotrophic primary production or microbial heterotrophic production in a given lake. It is one of the most important results of limnology that in most lakes phosphorous is the first limiting element. Nitrogen or carbon are further macro nutrients which have been observed to be limiting.

There is a general rule that the requirements of a lake for C:N:P is 105:16:1. Besides the macro nutrients, several organic and inorganic micro nutrients may limit the production. The following elements are essential for all algae: Fe, Mn, Zn, Co, Mo, Cu, B and V and the vitamins B-12 (cyanocobalamin), B-1 (thiamine), H (biotine) may be essential for some species. In principle, they are all candidates for the production limiting factor.

The amount of the substance which is actually available is, as in soil science, not necessarily identical with the amount which is present. Chelators, as humic acids, or the builders of detergents (polyphosphates or nitrilotriacetic acid [NTA]) usually influence largely the availability of all heavy metals thus mediating their nutritive or poisoning effect.

Elements fixed in organic material are not available for plants. They may be liberated in very different rates which range from turnover times of a few minutes - liberation of PO_4 through damaged membranes in the "short-circuited" P-cycle,- to almost eternity as applies for organic C which is immobilized in peat.

The considerations about the confinement of production as described until now are an over simplification. They imply - as the law of the minimum in classic plant physiology - that one particular factor limits growth. However, different algae have different nutrient requirements. The limiting substance is usually determined in bioassays with additional administration of potential confining nutrients. This definitely determines the limit of production in the present population, but if a water body is exposed to a varying pollution impact over a time long compared to the mean planktic generation period, the population may change and species with different requirements may become cominant.

Every element becomes toxic if it is concentrated above a

certain level. Under the conditions of surface water, Cu is the first candidate for this. It is reported to be poisonous at concentrations of 1 μg/l if not masked by chelators. Corresponding figures for other elements are: Mn, 200 μg/l for blue-green algae; Co, 2 μg/l; Zn,200 μg/l. Pesticide residues and their degradation products are among the substances which may limit production by poisonous effects too.

THE DRINKING WATER SUPPLY

Victor J. Kimm

It is a pleasure to be with you this afternoon and discuss some of our adventures in implementing the Safe Drinking Water Act. Bear in mind, that it is the primary duty of anyone who comes to the podium this late in the day to try and be brief and to the point, and so I will try to talk about some of the major provisions of the Safe Drinking Water Act. I will discuss the strategy we are using in trying to implement the bill, try to give you some idea of where we are with respect to specific steps, and then, perhaps respond to a few questions.

To begin with, the Safe Drinking Water Act is the newest of the environmental legislation that EPA is carrying out. It is public law 93.523. It's objective is fairly clearly stated. It includes a number of features that are somewhat different than previous environmental legislation, and I think those are the more interesting topics.

At any rate, the statute provides that the implementation, that is, the enforcement, will be primarily a state responsibility with EPA required to implement the regulations only in those instances in which the states either decline to accept primacy, or are unable to carry out the specific responsibilities.

Basically, there are two major programs created with this legislation. The first deals with regulations concerning the quality of the tap water, if you will. There are a number of requirements here that link together, and I will try and run through them briefly. To begin with, EPA is required to promulgate interim primary drinking water regulations. These are to include both maximum contaminant levels for specific constituents and, where necessary, treatment technologies in lieu of a maximum contaminant level.

The standards themselves are to be based on health protection, bearing in mind the costs - the full social and economic costs - of obtaining the standards. The Congress directed us to build our initial (interim) standards on the activities associated with the Interstate Carrier Program, which is a program we have had in the United States for a number of years, and with that program, has been the issuance and periodic updating of water quality standards of the Public Health Service. Those standards were most recently revised in 1962, and the current efforts of the interim standards are primarily based on the 1962 standards.

In addition to that, EPA is to enter into an agreement with the National Academy of Sciences, which is well underway, to take a look at the health effects record associated with the contaminants that we currently are regulating, as well as any other additional contaminants which they believe may be causing adverse effects. And this will be primarily a toxicological review; that is, dealing exclusively with the health effects and not bearing in mind the other factors which our legislation requires that we weigh before setting a standard.

As soon as this report is completed, ten days after it is turned over to us by the National Academy of Sciences, we are to publish it in total. And then, within 90 days, to revise our drinking water standards, which will then be the primary standards, to reflect any new evidence which has been gathered over this period of time. The time frame on that is such that it will occur in about two years. That is, we should have the revised standards by December of 1976.

The statute also provides us with the procedure for granting variances and exemptions. That is, in those instances in which the water supplier can demonstrate to the regulatory authority, which would be in the states in the cases in which they have assumed primacy, or to EPA, in the cases where the state has not taken on this role. There are procedures which would allow a stretchout in the compliance time. The justifications for that are either that the rural water quality is such that it is impossible to attain the standards, such as an isolated community with a single source. Secondly, where there are other good causes, primarily economic, to make rapid attainment of the standards infeasible. However, such variances and exemptions include with them a requirement that there be established a schedule over which the standards will, in fact, be attained, so that these are not unlimited outs. Finally, the statute provides for Federal grants to states to enable them to get some of the additional manpower and resources which will be necessary to implement the program. One feature of this legislation which is unique, and the source of something between major consternation and absolute fear, in the minds of the water suppliers, is a requirement of public notification. That is, the statute says that any time a utility either fails to undertake the monitoring required, or finds that they exceed one of the maximum contaminant levels in the standards, they must notify their customers. The implications of this may potentially be very serious, and I think probably will, and it is this anticipation of what may happen which I think is the source of so much consternation. If you look at the evolution of environmental legislation, this is really one more step in the direction of public participation and awareness. I think this pattern of citizen involvement and participation can be looked at historically, initially, in terms of the freedom of information legislation which has provided very clearly that citizens have a right to the documentation on which regulations have been established.

The law is concerned that citizens have a more direct access to the information with which they need to evaluate the reasonableness of the regulations that we have. The freedom of information laws have opened at least the Federal government and its documentation to the public. The current generation of so-called "sunshine laws" in the states have gone beyond that and have said that principal decision-making meetings need to be open to the public so that they may attend such meetings and hear the back-and-forth arguments. The citizens' right to participate at the Federal level is clearly established under the Administrative Procedure Act, which requires the publication of proposed rule making in hearings and public comment. In the case of the National Environmental Policy Act, the EIS process has clearly established the citizens' right to participate very fully in individual projects.

As far as the citizens' right to have information is concerned, this went one step further in 1972 when Congress began to include provisions which would enable a citizen to sue the administrator of an agency for failing to carry out his responsibilities. This is reflected in both the Clean Air Act and the Water Pollution Control Act.

In this context, one might look at the requirements for public notification as the next step, that is, giving the citizen the right to know when the standards are not attained. And, I think we will all have an interesting adventure, seeing how this turns out. It is my own hope and belief that if we are able to alert the public as to what may be coming so they can reasonably understand when the notifications go out the first time, that this will be a very effective way in alerting citizens to deficiencies in their utilities. Hopefully, these same citizens will support the necessary expenditures and the increased cost of the commodity necessary to bring about the attainment of the standards.

The second major program that has been instituted with the new bill deals with underground injection, and this is a very difficult area. The statute requires EPA to promulgate regulations to protect existing and future sources of underground waters. In addition, EPA is required to list the states for which such programs are necessary. In addition to that, once EPA lists the state as needing the program, the state then has 270 days to formulate an appropriate program. We have 90 days to review and either approve that state's program, in which case they take primacy, or to promulgate regulations under which we will administer the program in that particular state.

The House report makes it very clear that it is their expectation that all states will eventually be listed as needing a program. So, the flexibility we have is for some initial staging, should that prove necessary. The area that is of most concern to us in the underground injection is the breadth of the definition we use to define underground injection. That is, if one talks merely in terms of underground injection through deep wells, there are a finite number of such installations and most of those are currently under control. If one goes to the other extreme, you would require permits and case-by-case review and assessments for all of those practices which would endanger underground water which, as near as we have been able to determine, is virtually everything. The House report is quite explicit in this area and indicates that we would use these authorities to deal with almost all of the principal sources. The only exclusion was individual septic tanks. These involve problems that the agency is currently looking at.

We are taking a middle position in drafting the proposed regulations. The statute provides that some underground injection practices may be controlled by permit requiring case-by-case decisions, and others may be controlled by rule making, which is a general regulation of general applicability. We are trying to segregate underground injection practices into the most serious ones, which would require permits in other categories, like pits, ponds and lagoons, which the states, as they implement the program, will be able to initially deal with in terms for regulation. And then, as they obtain better information on what may be causing an environmental problem, switch from rule making to their option at a timeframe they find appropriate and consistent with the state's problems.

Let me turn briefly now to the strategy we are using to implement the bill. As is often the case with new programs, we begin by saying, what we need is a strategy so that we will be somewhat predictable and understandable to the real world, as to how we are going to implement the bill. Many of its responsibilities, of course, are very broad and not very precisely defined, so we produced a draft. It is still a draft.

We felt that we faced a serious problem in implementing this strategy and in implementing the bill at this particular point in time. We are very much concerned that we put together a program that is basically reasonable and can, in fact, be implemented. Thereby, we evolved a number of concerns and a number of priorities which are laid out in the document, and I might highlight them briefly here.

The heart of our strategy is to be as reasonable as possible and to get as many states as possible into the program early. The desire is to build on the capabilities that the states currently have in water supply. We look to a phased implementation, so everything is not tried the first year, but over time, develop effective and strong agencies at the state level.

Our basic program priorities deal with essentially a number of finite considerations. One, obviously public health, is our first concern. Secondly, since the problem is going to outstretch the resources by a considerable distance, we are very much concerned with setting priorities and dealing with the worst problems first. We intend to consider costs in all elements of the program. That is consistent with the statute and is fundamental in keeping the requirements reasonable. We will build to the maximum degree possible on the capabilities of the state level, and in fact, we are working hard to make the states partners in formulating the regulatory process.

Our strategy is based on the belief that public notification will work. In other words, an informed citizen will be an active supporter of those measures necessary to require the attainment of the standards.

Finally, we recognize the great difference in water supply problems - especially in the underground control programs area. There are great differences between and among the states so we do not expect that there will be one pattern for all state programs, which we will try to pressure states to accept. We are looking for substantive compliance with those particular activities spelled out in the statute and some others which we think are absolutely crucial in maintaining a program, and providing the state with as much flexibility as is consistent with the statutory requirements. We recognize that this will probably result in the states not adopting uniform structures. I think this is both necessary and reasonable.

Well, that is a kind of ambitious beginning. I began my tenure in this job at the end of May, and during the first week, went out and met with 47 representatives of states to show them just where we were. In reviewing the strategy document, I think it was very pleasantly received by the states. They were pleased to know that the program has flexibility where flexibility is appropriate, and also our interest in discussing with them and interested parties, including the industry, the environmentalists, and others concerned about the regulatory program.

We discussed with the states the interim primary drinking water regulations as they were proposed in March, and on which we had held public hearings. The states' concerns were primarily in the area of monitoring, though they felt the maximum contaminant levels were reasonable. They also felt, however, that the frequency of monitoring for a number of the parameters we had described, were unreasonable. About this same time, we got the first products back from an economic consultant working with us. He estimated that the monitoring costs initially included would amount to something like $150 million a year. We concluded, therefore, that this was unreasonable, so we revised the monitoring requirements and now have a program which, when implemented, will cost something like $30 million a year, which we believe is reasonable.

I think the statute is quite clear that the burden of having safe drinking water should be borne by the person who drinks it. That is, the utility. But, as we discovered talking with the states, both by practice, and in a number of states by statute, the states have, in the past, done the monitoring for small systems. They are very anxious to continue to do this in the future because they feel that is the only way they will get good service. That was a specific issue concerning the regulations we had in dealing with the state grants for the tap water program. On our initial draft, we said that routine monitoring would not be an eligible cost for a state grant after the second year. After hearing the states' discussion on this, we dropped it from the regulation as it was ultimately proposed. The public comment period on those regulations is rapidly coming to a close and, for the most part, the comments we have received on the state regulations have been very minor. We have extended the public comment period at the request of the NRDC, and I believe that their concern deals with more requirements for public participation in the formulation of a state plan. We still have not seen that.

With respect to the underground injection control regulations, the first draft of these were worked on at the early meetings in June. That first draft included a very broad definition of potential sources of underground injection, and basically, required permitting all of the different types of activities. Here again, the states pointed out to us that the associated work load to do in a year or two was absolutely prohibitive, so we adopted the mix strategy, as I described earlier - mixing, permiting and rule making. But over time, we will still hope to bring under control a variety of practices that may be contaminating and are contaminating our ground waters in some instances.

In connection with all of this regulatory work, we have had, perhaps for the first time in EPA's history, a large number of meetings with states and other interested parties. Although at times I think I may be two or three years ahead of myself in listening, it has been a very effective process and one I think is going to help in the quality of our regulations. However, that type of an open process takes time, and the statute said that by mid-June we should have issued a number of standards and regulations which are now just coming into the Federal Register and will appear within the next few weeks. On balance, my own judgment is that time and the opportunity it has afforded us to talk with people who will be directly impacted by the program has been very beneficial. Although I have received two notes of intention to sue, which is required by our statute, no one has yet dragged me off to a district court to have me presented with a mandatory schedule of when we will get regulations out. I said - not yet!

Finally, in conclusion, I think at EPA we are getting together a basically reasonable program and a program which will in fact be supported by a number of the states. To be realistic, we face very large problems. To begin with, there probably was not a worse time in the last decade to begin a new environmental program. The Federal budget is tight and state budgets are extremely tight. Of the institutions that I have been talking with, primarily state health departments, a vast majority of them are facing either an absolute ceiling, which with inflation means a smaller program, or a direct cut in resources. One state in particular is facing a 28% cut in resources, which means they are going to have to let staff go. Budget constraint is going to be a very serious problem for this program.

There is also a great deal of cynicism and skepticism, and especially among state legislatures, that the Federal government will start an ambitious program, limit the amount of grant assistance it provides, and basically leave the states with a large new program which they are responsible for implementing.

Finally, there are a number of areas we are dealing with in drinking water, and I particularly have in mind the significance of small quantities of organic contaminants, on which the scientific evidence is very preliminary and we are, I think, quite truly at the state of the art of what anybody knows, and about which good people can disagree very vehemently about how protective one ought to be. Hopefully, with the resources that we have received for research, we will be able to extend our knowledge in this area. But as many of you may have noticed in the papers, last week a firm found a new analyzer. With the new analyzer, they were able to determine the presence of a family of organic compounds, nitrosamines. And sure enough, they checked a dozen water samples, primarily in the New Orleans area, and in fact, did discover the presence of nitrosamines in very dilute concentrations. The significance of these is very difficult to evaluate, and also the kind of ongoing problem the agency will have to deal with in this media as well as others.

To conclude then, I am reasonably optimistic that we are going to be able to get the program together. From my interactions with the states, it would appear that a number of states are going to work with us, that is, to move towards primacy. I think if we can get a collective effort, we can, in fact, make significant progress towards improving the quality of the nation's water.

DISCUSSION

Dr. TRUHAUT: Dr. Kühn, are you aware of investigations made in the Institute of Public Health of the Netherlands by a team using organisms as a means of analyzing the Rhine waters, the waters of the estuary containing concentrating organisms of the algae family? They found that about more than 1,000 chemicals were concentrated between a thousand and ten thousand fold, the concentration in the water of the Rhine. Are you aware of that?

DR. KÜHN: Yes, we especially are not working in looking for the contents of biological material. But I agree with you that a lot of the organic chloro compounds are summed up in the nitrification curves, in the algae and fish. In our Institute, we determine also the content of these compounds in some fish, and in some other animals, such as ducks, living on the River Rhine. Therefore, I would say it is not too important to speak about concentrations and to speak about ppm or ppb in the water itself, because a lot of the toxic material is summed up in the body fat either of animals or of man. Therefore, I would not agree with Paraclesus, because we do not have to watch the concentration in the water. It is much more important to watch the concentration in the effluent after the sand bank filtration or the concentration in our own body.

DR. TRUHAUT: By the way, the concentration is not always a sign of toxicity. For example, if they are living organisms, for example, crustaceans, which are able to concentrate the arsenic of the sea water, the concentration in this case is a kind of detoxification. For example, in the lobster, there is an amount of arsenic which is near 40 ppm. At a first glance, a hygienist looking at the lobster would have said, "Don't eat lobster, please. There is too much arsenic in it." And it is not true. The arsenic is detoxified. Poulson and his collaborators made a very spectacular experiment where they gave rats in their diet certain concentrations of arsenic, and also put the same amount of arsenic contained in shrimp as another diet to a different group. The rats eating the shrimp were in very good shape, not only acutely, but chronically, during their whole life span. On the contrary, the rats submitted to the same quantity of non-shrimp arsenic gave signs of toxicity very early, sub-acutely, at least. This means that in certain cases, the concentration of a given element or substance in a given organism is a kind of detoxification and not a toxification. The concentration means not always toxicity. I hope I am clear in my broken English.

DR. COULSTON: In the case of arsenic, it is pretty well established now that the penta-valent arsenic, which is commonly found in most shrimp, is not as toxic as the tri-valent arsenic, which is the common arsenic that you are talking about. So what Dr. Truhaut said is correct. This study of Dr. Poulson in Denmark was a very clear experiment. He was trying to demonstrate that relative amounts of metallic arsenic, forgetting what valence it had, didn't really matter, and that there was a distinct difference between the ability of the animal body to handle it in its different forms.

DR. EGAN: May I say, this also tends to be an analytical problem. Of course, it does need to be clarified in the way that we are now discussing by the toxicologists.

DR. vanRAALTE: I will be brief, but maybe a little belated, in fact, 450 years too late. I have got to come to the support of Paracelsus when he said the dose only. It is only the dose that makes the poison. Because I want to remind all of you that it is the dose that determines the concentration in the organism. And what Paracelsus said and is still true. It applies to the species and to the total intake and to the concentration in the target organ. And, therefore, it is just as true today as it was in older times.

DR. EGAN: I think Professor vanRaalte is quite right in coming to the defense of Paracelsus. It depends upon what you are talking about, and we were talking about two different things, tending to. But, in any case, the dose is what matters.

MR. CLARK: In January of this year, there was an article in Science, using Carbon 13, Carbon 12, ratios as an indication of the amount of the dissolved organic carbon in rivers that was due to the influence of man, that is derived from petroleum, basically. Are you using these methods? The Potomac River, for example, was shown to have 40% of its dissolved organic carbon due to petroleum derivative sources of carbon. Are you finding similar numbers for the Rhine?

DR. KÜHN: Yes. We used this method in earlier years, but we don't use them now. For example, at the River Rhine, in some industrial cases, we have a content of 70 and 80 percent of organic chloro compounds, and I see no way that this 80% of the compounds could be produced by nature.

DR. COULSTON: Dr. Kühn, I am a little confused. I hear this talk about the Rhine River being so heavily polluted, and yet, I know that many great cities along the Rhine take the water from the Rhine to use for drinking purposes. How do they clean up the water? What is the mystery? In this country, we tend to believe you cannot take Hudson River water, or Potomac water, and make it potable water - drinking water. But yet, the Rhine _must_ provide drinking water. What is the difference?

DR. KÜHN: I spoke about two types of standards: the Standards A and Standards B. At the River Rhine, we have had for over twenty years the Standards A, that means only sand bank filtration. It is sure impossible to clean all the Rhine River water up to now if you have our standards and our maximum data measured as we measure them today. We have the river sand bank filtration. It was enough, but it is not enough now. Now, we have the floculation, ozonation, adsorption and disinfection techniques, and these are enough for this highly polluted river. But, there is a breakthrough of a lot of substances, and we think, if the condition of the River Rhine would worsen in the future, we couldn't treat the water with the possibilities we have now. And, we think that other possibilities, probably distillation or something, is not the right way to produce drinking water. We think it would be better to keep these standards or to make the River Rhine a little cleaner.

MR. KIMM: Are you doing work on the reaction kinetics for the formation of some of these organic groupings?

DR. KÜHN: Yes, we are working in this, and we looked for the precursors of the haloforms, etc. The work is just in the beginnings of research; the EPA is working in this field, and some humic precursors from chloroform and tetrachloroethylene have been found.

MR. KIMM: I take it, then, you are generally sympathetic with the approaches that others are using - they are taking a very careful look at humic acid as the basic building block, if you will. But, have you characterized some of your ozonated supplies to find out if you do find chloroform in some of the other haloethers and halomethanes?

DR. KÜHN: We are working in ozonation, and, you know, in the beginning, ozone was introduced at the River Rhine to take the manganese 2 out of the water. It was oxidized to permanganate and reduced later on by the activated carbon. We figured out that there is what we call a microfloculation during the ozonation. That means, the floculation is working much better after ozonation, and there was also a better removal in the activated carbon filters. But that is hard to understand, because during ozonation, you make much more polar substances, and polar substances are not as well absorbed as non-polar substances. But, in the activated carbon filters, the substance is better biodegradable after ozonation. A lot of people are scared, and someone told me, in the United States, "I made some research with ozone, but there was much more biological activity after ozonation." And I said, "Yes, that's right, and that is very nice. We want to go on this slope of the curve, that the substances become more and more degradable." We are working in this field to determine what compounds are produced during ozonation with a high liquid pressure chromatograph, but we cannot give an end result in this, yet.

MR. KIMM: You have indicated that you have some supplies that currently use carbon absorption. And you said there were at least some contaminants, there appears to be fairly rapid breakthroughs. Could you identify some of the organic families that seem to be coming through the carbon quickest?

DR. KÜHN: Yes, the polar organic compounds, like humic acids, are coming through very quickly. If I go back in history, the waterworks at the River Rhine had carbon filters running for one and two years to remove taste and odor. They analyzed the water and filters. They looked for the COD breakthrough, and they had to remove the carbon every six or eight months. Now, we are looking for a breakthrough of organic chloro compounds and the waterworks have to generate the carbon every three or four weeks. But the compounds coming in the beginning are the very polar compounds, particularly the compounds with the high molecular weight, humic acids. But we are not very scared about the humic acids. I think it is not possible to take all the organics out of the water, and we also need some organics against corrosion. But, we are looking now for the organic chloro compounds, and I think this is a group breaking through after the very polar compounds.

MR. BUCKLEY: I would like to make one comment and ask one question. The comment relates to the fact that we have to be very careful about not oversimplifying this thing about dealing with chlorine compounds. After all, there are a lot of chlorine compounds around. Many of them are inorganic. There are organic compounds produced which are, for example, antibiotics, and I think we have to make a distinction between some of the short chain hydrocarbons and some of the ring compounds which are chlorine substituted in the process of chlorination. As compared with some of the other materials which are present in municipal effluents and industrial effluents, which are large molecules which may pick up an odd chlorine compound, and may be picked up in a gross analysis, and some of these materials are, let's say, breakdowns from chlorination, bleaching. You got a different set of compounds coming from craft mills. From sulfites, you have another series. These start out originally as very large molecules. They are broken down into still fairly sizable fragments, and they do pick up an occasional chlorine molecule. But, there is just no evidence to suggest there is any relationship between these and the build-up of chlorinated hydrocarbons in fatty tissue. I think we have go to keep our minds open - not pre-judge some of these situations here. For example, one of our first interests in chlorine in water, is its relationship to odor and taste. We could detect compounds, chlorophenols, particularly, in just a few parts per billion, and really, just focus our whole attention originally on it, as far as municipal water was concerned.

Now, we picked these up from road tar washings and things like that that got into our water supplies. There is a whole family of chlorinated compounds out there in surface waters, many of them derived from just a decomposition of organic matter. If this gets into a municipal system, which is post-treated with chlorine, we are going to end up with a whole family of chlorine compounds. I just urge you not to draw any broad conclusions until we have a lot more data on it. We have some specific data on some short chain hydrocarbons. Let's not extrapolate this to infinity at this point in time.

The question I would like to pose, is wherever indeed on the Rhine you do have post-chlorination of municipal wastes? Because this is a fairly common practice in this country, and we are starting to ask some serious questions about the wisdom of all this, and we would ask what is your experience with municipal waste water supply.

DR. KÜHN: I want to quote my supervisor. He talked a couple of months ago in a meeting, and he said, "Chlorination in the drinking water field is dangerous, but in the waste water field, it is criminal."

MR. CLARK: Mr. Hutton, you did call for greater public health effects, experimentation and research with regard to bottled water. If I could ask these as a series of questions: can public health effects experimentation be carried out on man? If not, then would you accept animal research? And if so, what factor of safety would you take in applying the results to man? This seems to be a very critical issue in this meeting. I would like Mr. Hutton to comment.

DR. HUTTON: I am afraid that there are some toxicologists here that might be much better prepared to take a shot at that question than I am. The only point that I would care to underscore is that if we continue this headlong pursuit of this constantly receding analytical zero, I am concerned that we are going to find ourselves in a hole that we cannot afford to buy our way out of.

DR. TRUHAUT: This is really a fundamental question relating to toxicological evaluation. And what margin of safety is to be applied. As you know, the current policy is to apply to toxicological evaluation of a certain chemical, a safety factor of one hundred, for extrapolation to man of the data from the most sensitive species. This is to be on the side of safety. It would be better to establish what is the most suitable, rather than the most sensitive species. What is the animal species in the laboratory which is closest to man? To understand that, you have to remember that this safety factor of one hundred is applied on the basis of Arnold Lehman's and Alistair Fraizer's work. On what basis? First of all, they said that if we don't know the sensitivity of a given animal species comparatively to man, we have to suppose on the safety side that man is the most sensitive animal, and apply for a factor of 100 for the variation of sensitivity between species.

Second, in the case of laboratory animals, we are making experiments in the laboratory on homogeneous lots of animals. That is, they are very often of a given strain, of a precise breeding, and for this reason, we have to apply another safety factor, because it is not the case for human populations, which are very heterogeneous. They include children, pregnant women, old people, ill people, adults who don't know they are ill, but have some impairment, and so on. And for this reason, we have to apply, as was the theory of Lehman and Fraizer, another factor of ten, and ten multiplied by ten, this gives one hundred. Of course, this must not be, if I may say so, a biblical value, that is, something very rigorous, such as a mathematical value. And, for this reason, in those organizations such as WHO, FAO and others, sometimes the safety factor is decreased and sometimes it is increased, on a basis of value judgment. This is the situation. But, some day, I hope, because of the work of toxicologists, and especially of related disciplines like, for example, biochemistry and molecular biology, we will be in a position to find the most suitable species. This means that every case has to be judged on its own merits by competent people. And by competent people, I mean competent experts. I have in mind not only those who are knowledgable, but those who, in my view, have a good sense, a good value judgment. It is even better than the knowledge, because in toxicological evaluation, you have three stages. First, you have to obtain results by studying the toxicity of a given substance, qualitatively and quantitatively. These results, as obtained by experimentation in laboratory animals, and, as much as possible, by observation in man.

The second stage is to interpret the findings, the results in animal laboratories, the results of observation in man. You might think that we are in better state than twenty years ago, because of the simple fact that the methodological approaches for toxicological evaluation have been very much improved. We have now electronic microscopy, biochemistry and enzymatic induction, etc., but paradoxically, the interpretation of the result is more difficult than it was in the past, because you have many more parameters to be considered.

And now we come to the third stage. And here the toxicologist must not be alone. The third stage is action to be taken, facing the results and their interpretation. And for this, the toxicologist must be there. I will now speak about, or refer to, the famous concept of Benefit versus Risk. To evaluate risk, you have to have toxicologists, of course. To evaluate benefits, you have to sit with economists, decision makers, the decision must be made after taking into account the benefits for the community. For example, if something is really improving the welfare of the population, it would be completely silly to exaggerate the risks, if there is a great benefit for the welfare of the population. Each substance, in my view, has to be evaluated on its own merit, in referring to a number of parameters which can now be obtained more easily by the modern techniques of science, research, and so on.

DR. ANGELOTTI: I would like to make two comments which would be connected with Mr. Clark's remarks. Let me speak momentarily to the enunciated cancer policy of the United States. If it is not officially enunciated, it is, at least, unofficially enunciated. And that is that, as national interests, we are attempting to eliminate carcinogens from the environment wherever we can find them, and with a speed that is commensurate with our abilities. Keeping this in mind, let's next look at one of the legal complications that we have in our country. That is that water, under the Food, Drug and Cosmetic Act, is defined as food; therefore, water is subject to all of the requirements of the Food, Drug and Cosmetic Act. The Food, Drug and Cosmetic Act has a food additive amendment that was written into it in 1958, and to which was added the Delaney Clause, which, in essence, states that no carcinogens may be added to food. Now, if, through our current municipal water treatment system, inadvertently or otherwise, we are introducing carcinogens, this presents a problem for us nationally. For when someone identifies to the Food and Drug Administration that a carcinogen has been produced in a treated water, and that water is being used for the manufacture of foods and drugs, or is being taken in as food per se, as in the case of bottled water, the Agency has no course, under the law, but to prohibit the use of that water for those purposes. Now, we do have some options open to us as an agency. Instead of banning the use of the water, we could require a treatment on the premises to remove that particular carcinogen. So the difficulty that we have before us nationally is what are we to do with the Delaney situation? As long as it exists on the books, it is a reality. Whether it represents good science or not has nothing to do with the issue. The Delaney Clause does not allow for a consideration of dose response, the very thing that Dr. Truhaut was explaining to us a moment ago. And, in the absence of a dose response consideration, it is very difficult to deal with any rationale when one is attempting to do something about carcinogens in the environment.

DR. SMEETS: I don't want particularly to refer to the definition of an epidemiologist, or a toxicologist, as I recently heard, that those are people who are trying to make the world liveable for rats and mice, but this is a point apart.

I listened with great interest to Dr. Hutton's reasons for the consumers to want bottled water. And he was talking about better taste, uniformity, safety, and health. But at the same time, he stated that all minerals are removed from bottled water. This brings me to the first comment, as I already mentioned earlier, we had a colloquium, a scientific meeting on the hardness of drinking water and cardiovascular diseases. And one of the most important conclusions we made on this particular point is, that in the current state of knowledge, artificially softening drinking water cannot be encouraged. I wanted to know, what are the reasons for you to say, Dr. Hutton, that distilled water is healthy, and that it is safe.

This brings me, at the same time, to a second point, that we, from the Commission, as I will inform you in my upcoming presentation, have introduced in our proposal for drinking water standards, which are now in the Council of Ministers, not only maximum allowable concentrations, but also a certain number of pollutants, minimum allowable concentrations. I am referring particularly to calcium, magnesium, and some others. Since we have your opinion that there must be a particular concentration of those materials, this brings me to yet another question. I address myself at the same time to Mr. Kimm, and also to the representative of the Food and Drug Administration, establishing of drinking water norms has to be in a collaboration between people of the EPA and FDA with regard to the norms. Since for me there is still a great question for which I cannot get the answer, that is, when there is a lack of special minerals or trace metals in drinking water, are there any normal standards with food. Or, I can say it in another way. When there is a large concentration of minerals or elements in drinking water, there is at the same time a maximum allowable concentration in food. When you put those both together, what then is the situation? Has there been any collaboration between the drinking water people of EPA and the Food and Drug Administration, in establishing this combined food-water norm?

But first, I would like to know from Dr. Hutton how he knows that distilled water is completely safe and healthy?

DR. HUTTON. Evidently, I misstated the situation, because the demineralization, as we practice it, is simply the first step. We do sell a demineralized product, but that is for steam irons and that sort of thing, not for drinking. The water which we sell for drinking is re-mineralized. As a matter of fact, we add back both calcium and magnesium.

DR. MRAK: I wanted to mention that the Delaney Clause talks about an appropriate test. We tried to get the FDA to realize that, and the Commissioner was willing to do it, but it was stopped by the lawyers.

DR. COULSTON: I think it would be better to ask what knowledge we have about the required total intake, because drinking water is only one source of intake. We get additional amounts of minerals in food, by inhalation, and even by smoking. And it is the total intake which is decisive. Minerals in drinking water are a good indication of local factors, in cases where people drink local water. Data about these minerals may be useful in understanding if the intake is excessive or if it is not sufficient, what relationships they have to disease processes, in general.

MR. JONES: I am with the American Bottled Water Association, I think the only association in the United States that concentrates on nothing but drinking water. I would like to respond to the questions that have been going around about the fact of taking minerals out of water and then putting them back in. Remember that Dr. Hutton is speaking for one company of our Association. We do produce distilled water, and we do sell it, but we sell it to people who want distilled water, because their doctors have told them to drink water without minerals in it. I will also respond to the questions that came up earlier today, about fluoride in water. Now, we, too, provide fluoride in water, in drinking water, if people want it. We do not promote it. We will say that if you feel your child should have it, we have fluoride water for him. Interestingly enough, we were concerned, and you may be interested, in whether or not our sales went up or down when fluoride was added to a city water, and we have checked that in three different municipalities, and we cannot find any difference. Those people who did not want it, did not buy it. And the balance of our sales in those municipalities stayed exactly the same. We still have the comment, from people calling in on us, that it was a foreign plot to put fluoride in the water, and they wanted an explanation from us as to why we didn't object to it.

Now, many of our companies take the minerals out of the water and then put minerals back in to taste. The quantities of minerals that go back in are definitely regulated in their number by the FDA, since we come under them as a food. The amount of each mineral is definitely regulated by them in the established standards. The actual minerals that are picked and are put back into this re-manufactured water will vary with each company, and in many cases, that is as closely guarded a secret as the formula for Coca-Cola®. I have been at a number of tastings that water companies have had, to determine what is the most palatable water for the particular municipality, or that particular sales area. And I have always felt that if someone was born in White Sulfur Springs, Texarkana, and the water tasted slightly of sulfur from the time he was born, that, to him was good water. So, we have to sell our product, and it is taste on which we sell it. So, each company will vary it on that basis.

I would like to comment on a few other things that have occurred. Dr. Angelotti mentioned that we petitioned the FDA for good manufacturing practices. We did, and many of my peers have wondered why we took that step, because there are enough government regulations as it is. Why would we even ask for them to get involved? We asked for them to get involved for a number of reasons.

We represent about 93 to 95 percent of the bottled water business that is done in the United States. We are a large organization. We have some good established companies, including Foremost, McKesson, Nestles, Coca-Cola, and Borden. We also have some people who are not in our industry, some of the smaller companies. And we felt that it was imperative that there be regulations over our industry that could be controlled by someone other than ourselves. We do a self-policing job, and we think we do a very good one. We have established guidelines long before they were put upon us. We hired outside research organizations. First, Syracuse University Research Corporation and, lately, the American Sanitation Institute, to go around to our plants and inspect every single one of them, to see that the quality of the product met regulations, and to see that the plants were at the state of sanitation that was needed to produce the best quality of water. We check the plants constantly and help improve them. But more than that, we were concerned about someone who was not in our industry, who was not being tested by us, who could not be thrown out of membership if they didn't comply, who did not come up to these standards. That is why we went to the FDA and asked them for some standards to be established upon us. Now that we have them, we are doing everything that we possibly can, not only for our members, but for non-members as well, with schools and training courses, everything we can do to encourage them to bring their standards up-to-date. We have been very pleased, and we will work with anything that you want to come up with to improve the quality of the bottled water that we produce.

And that brings me up to the big question. What quality of bottled water do we produce?

When a situation occurred in New Orleans, and we had the carcinogens in water explained to us out of a clear blue sky, my phone rang like mad. The phones of all bottled water companies throughout the United States rang like mad. We would like to see perhaps a little more restraint, or a little more careful briefing of people, when articles are released to the newspapers, because it was said that it was not sure that these chemicals were harmful. The public certainly got the impression that they were harmful, and we were in a lot of trouble. Is there any way that this can be helped?

MR. ANGELOTTI: I would merely say that yes, there is cooperation between EPA and FDA in the establishment of the drinking water standards. EPA has distributed, and will continue to distribute, their proposed standards for comment from relevant sister agencies at the federal level, as well as from the appropriate folks at other than the federal level. And in our review of the proposals made by EPA, we do take into consideration the concept of total body burden, particularly where the heavy metals are concerned, that is part of the consideration, yes.

DR. KORTE: Dr. Ernst, if I understood you corrently, you mentioned the acid rain which produced some acid lakes in Sweden. I learned from a couple of publications that the SO_2

emission from the industrial coal district in Essen, which has been since and before the First World War, a distance of about 150 kilometers. That means, that it could never reach Sweden. I would like your opinion on this.

DR. ERNST: I did not say it is the German industry, I said it is the Middle European. I referred to several articles, short reports, which I do not definitely remember in detail. One of the reports even said by some tracer analysis that it had been established that at least the bulk of this did not come from Germany. My general recollection concerning these articles was that no one was really sure where it did come from, if it was a worldwide distribution or not. So, I would like to stress that the Swedish Lakes are acid, but I did not say what directly causes this.

DR. COULSTON: I heard the Deputy Administrator of EPA give a television press conference, where he stated very clearly that the spread of emission to bodies of water is a very great problem that has to be investigated very thoroughly. This is particularly important, since he reported nitrosamines in some air samples.

DR. KIMM: There is a great deal of confusion about this. In the paper presented in Las Vegas, they reported basically two things. One, they had a new analyzer which enabled them to go to lower levels of concentration than had previously been reported, and they now had an ability to deal with a non-volatile fraction. They then reported that they had checked a finite number of air samples, as well as some water samples. Mr. Quarl's remarks were directed primarily at, as far as we know, the first recent finding of nitrosamines in air, which is more significant in that although the gross quantities were small, it was dimethyl nitrosamine, which is a very serious business. What he said was that the whole question of atmospheric chemistry is one in which we need to do more work. The remarks that I made were in another part of that paper, but as I noted earlier, this just indicates the kind of ongoing problem we face as better analytical methods come along. As we find very small quantities of materials which we have some reason to believe affects health, we will have to take these problems seriously, although the specifics of how you deal with them, especially in a regulatory framework, is very troublesome.

DR. EGAN: There is a paradox here. I would fully subscribe to the need not to go on endlessly inventing more sensitive analytical methods, but there is also the matter of identifying positively those residue traces that you do detect. And the question of drawing conclusions at that point is another matter, too.

DR. TRUHAUT: About your remarks, Dr. Kimm, about the significance of what I call micro-organic pollutants. I have been worried about that for a long time. I elaborated a project of research which was to test in the long term extracts from drinking water in France, polluted from the beginning, and treated. I was worried not only about the treatment by chlorine, especially, but also by the treatment by ozone. My collaborators and I tested in a project supported at the same time by the Ministry of Environment in France and the European Economic Community, the extracts of two kinds of drinking water: one treated by ozone, and the other, an extraction made by another team of analysts with chloroform at pH7. I was, at the beginning, surprised to see that the extract was like a tarry residue. This was given to rats and mice, of course with controls, two kinds of controls - a control without this water and a control with extract only from the water itself. The extract was given at several doses to produce a large safety factor. This means that the amount given to each animal, either rats or mice, was at least 100 times more than the consumption of man. And the results were bad. Because I don't want to create a phobia - I said this morning why, because this is not good to create fear in the public - this is why I have waited. Because I did not want to create a phobia, I did not publish this. I did not, it is only in secret reports. But the fact is, that the results were bad. In mice and rats, at the doses given, we obtained hepatic lesions, lesions of the adrenals, and lesions of the kidney. And overall, and I will not elaborate more about that, tumors. In the case of the mice, tumors, a very significant increase in the frequency of tumors, comparatively to the control. In mice, it was mammary adenocarcinoma. But in rats, females, there were tumors of the central nervous system, and in males, there were tumors of the kidney and tumors of the testes. At present, of course, it is very difficult to interpret.

DR. COULSTON: What do you mean by tumors?

DR. TRUHAUT: Malignant.

DR. COULSTON: You mean cancer?

DR. TRUHAUT: There were malignant and benign tumors. I can not give to you the histology data, you see. The main thing is that the results were really surprising, and this means that in some cases, i don't know what will happen with other types of polluted water treated to make drinking water.

DR. COULSTON: Just for the record, then, since you have said this, did you find these tumors in rats as well as mice?

DR. TRUHAUT: It was mammary adenocarcinoma in the mice.

DR. COULSTON: And in the liver of the mice?

DR. TRUHAUT: There were no tumors in the livers of the mice. In the rats, the females mainly, tumors of the central nervous system, tumors of the brain, and in the males, tumors of the kidney and tumors of the testes. This is all. It is very difficult to interpret this. Of course, I asked for an analysis of the chemical composition of the extracts, and this is underway. But you understand why I don't want to publish this, because to say to the people that they are being poisoned by the French waterworks in this case, would be, in my view, a kind of crime. Because the toxicologist has to ensure safety, but not to cry that people are poisoned.

DR. COULSTON: Rene, if I may say, please do it with water, but don't do it with French wine, please. It is too precious to all of us.

DR. TRUHAUT: I will defend the French wine, because in another case I have to test wine which was kept in a kind of tank with a certain kind of coating. I kept the wine in it during a certain number of months, even for years, and I had to give to the controls - of course, I could not give the wine itself, because I had to evaporate the alcohol first - but the fact was that the controls treated by the extract of wine from which the alcohol was eliminated, these controls had a very long life span compared to the other models. It is better to drink wine than water.

DR. EGAN: There is a moral here, which, again, is off the record, and that is - never drink from a wet glass.

I think we are grateful to Dr. Truhaut for candidly drawing our attention to some of the limitations of the conclusions which ought to be drawn from certain data at this stage, but I think a more detailed discussion of this might be more profitable individually.

In closing, I would like to thank all the afternoon's speakers and participants.

WATER QUALITY - PROBLEMS, SOLUTIONS, MANAGEMENT, STANDARDS, AND VISION

E. H. Blair

From the beginning of time, water has been both man's benefactor and his curse. The seasonal flooding of the Nile provided early civilizations with food and fiber, yet was a ravaging source of pestilence. For centuries, bodies of water and rivers determined the location of populations of the world, and the course of commerce and trade. Boundaries of nations and cities follow the meanderings of rivers. Fights for local property rights have perpetually focused on the oasis and the water hole. The industrial revolution standardized around the location of water and the feasibility of diversions of water. The high standard of civilization of Scandinavia is largely a result of hydroelectric power.

Even during this century, the perspective on water has changed dramatically. The dramatic problems of water in the early parts of this century concerned floods, typhoid, solid water in the wrong place - the Titanic, and diversions of tremendous quantities of water to enhance the agricultural capability of arid regions. Since the early 1960's, the problems have been more often:

Do we have enough?

How to desalinate brackish water?

How do we convert our rivers from the cesspools of our municipalities?

How do we control catastrophic spills?

Concern for latent health effects.

Concern for the effect of thermal pollution on wildlife.

And many more.

Summing up, in the context of history, water quality has been viewed in the perspective of where, when, diversion, new uses, harnessing it, and a whole host of other value assessments. Thus, it seems that this symposium is focusing on an important _concept of quality_, i.e., largely the permeations of water purity.

As a research chemist by training, I'm a lot more comfortable talking about solutions to _problems_ than about _universal standards and control_. Currently, as a manager, I find that solutions are still my forte and that motivation, education and research are the most effective tools in developing solutions to complex problems. I am therefore going to illustrate and emphasize some of the problems, and varied solutions, and management approaches to better purity of the water effluents of the industrial sector. Admittedly, I'll be slanting my remarks toward chemicals. And admittedly I will be using examples drawn from my association with The Dow Chemical Company -- an association approaching a quarter of a century.

First, let's look at the broad problem of chemicals in water. Sources run the gambit from natural sources; to agricultural runoff; to mining leachates, to effluents from chemical manufacturing; to effluents from a whole host of processing industries; such as textiles, food processing and metal plating, to transportation incidents, to municipal wastes. Within the chemical industry, our problems largely center around effluent quality, thermal effluents, and transportation incidents.

Considering chemical plants, we, at Dow, have had to develop three distinct solutions, somewhat predicated upon geography. Our Midland, Michigan plant is among the largest chemical complexes at a single site anywhere in the world. Yet, it is located on a very small river. For this reason, the plant has been a pioneer in developing a number of treatment processes, including pioneering work on biooxidation of phenolic wastes. Today, this plant, with a daily outfall of about 40 million gallons, includes a sophisticated tertiary treatment plant.

We also have large chemical complexes on the Mississippi River in Louisiana and the Brazos River in Texas. These plants, started in the 1940's and 1950's, were consistent with the chemical industry on the Gulf Coast at that time -- little treatment and direct discharge to the river. As the context of pollution concerns has evolved, both of these plants are using or are going to use secondary water treatment, utilizing much of the technology developed at our Michigan plant.

We have been able to develop another approach to pollution control at our plant in Pittsburg, California. Sunny California permits the ideal "zero" discharge plant. Evaporation rates are high enough (up to about 50 inches per year) and rainfall is low enough (10 to 15 inches per year) so that large holding ponds are feasible. This is not without it's penalty, inasmuch as long-term we may have a landfill situation to cope with.

I believe it is important to emphasize that several factors motivated the research, technology development and capital spending to achieve these accomplishments. Obviously, on a small river, phenol taste in sport fish downstream is a persuading factor to get moving. Many of our emission reductions have been accomplished in the <u>chemical processing part of the plants</u> - not by tail-end control technology. Here, the motivating factor has been the calculatable economics of not wasting materials. Finally, anticipation of standards has been another motivator.

Recently, ten companies composing a major segment of the chemical industry documented the degree of improvement in their pollutant discharges. Levels of BOD in 1974 were 50% below 1970 levels. By 1977, BOD, COD, and total suspended solids will be 5 to 20% of the 1970 discharges.

The annual cost of achieving the 1977 performance will be about $400 million. Now, by 1983, the U.S. Water Pollution Control Law stipulates that levels of BOD, COD, and total suspended solids should be a little lower. We are locked into a goal which is equivalent to 2 to 10% of the 1970 levels. To get from the 5% to the 2% level of BOD (and related minor improvements) will impose additional annual costs of about $300 million. Most reasonable men are concenred about the doubling of annual operating costs, capital investment in hardware, and energy consumption to go just a little further in discharge quality. Whether such reasoned judgment is politically viable enough to change an idealistic law is our current problem needing a solution.

Chemical environmental incidents have been a plague of the chemical industry for years. Endosulfan in the Rhine, chloroform in the Mississippi River, caustic in our Tittabawasse River, and cyanide from electroplating operations have all received headlines. Engineering ingenuity and management have provided the solutions. Engineering management practices run the gambit from diking around tankage, to sophisticated level alarms, to double interlocking valve systems on tankage, to extensive worker education and motivation.

Transportation incidents will occur. Without casting stones, there seems to be practical realities to the quality of railbeds, the limits of barge transportation, highway accidents, human errors in loading, and the mechanics of transportation. Our industry, through the Manufacturing Chemists Association, has taken the realistic approach that accidents will happen. We have, therefore, geared ourselves to have expert people available and to get immediately to the scene of an accident, both via telephone and airplane. Through cooperation of participating companies, teams of experts are positioned throughout the United States. Bills of lading contain an emergency response telephone number. Firemen, public health authorities, and local agencies call on the emergency response teams, who can immediately give phone advice and fly to the site to work with the local authorities and transportation officials dealing with the situation.

But really, an ounce of prevention is always worth at least a pound of cure. So it is with the transportation of chemicals. Our company is building a new terminal and warehouse to replace an out-dated facility. It happens to be on a fairly large river where a considerable amount of industrial traffic is currently taking place. The river is also used as a potable water supply for a number of cities.

Some of the planning, besides all the normal business considerations, includes the following engineering design criteria:

1) On-Land
   a) There will be no discharge of waste to the river.
   b) Spills on land will be contained by impervious diking at the site.
   c) The lay-out will be designed so spills can be recovered and appropriately disposed of, off-site.

2) The Dock
   a) Containment of spills on the dock by retaining curbs, and aboard ship by keeping deck scuppers closed to avoid drainage into the river.
   b) Maintenance of unloading hoses in good condition by frequent checks and replacements when needed to avoid spills between the ship and the dock by hose failures. Experience has shown that hose failures mostly occur near each end, which further reduces the likelihood of a spill between the ship and the dock.
   c) Use of floating booms around the ship when lighter-than-water products are being handled.

3) Ship Collisions

Chemicals will be moved into this terminal by ship, not by barge. Now, we have never had a major accidental loss of chemicals from ships in our 20-year history of ship transportation. However, we felt the greatest hazard of the new terminal is the potential for accidental contamination of a water body used as a potable water supply resulting from a ship collision. In anticipation that such an accident is possible, the following comprehensive emergency response plan is being worked out:

   a) Identification of potable water users and their alternatives if river contamination became hazardous, i.e., alternate supply, storage capacity, consumption reduction, water treatment, etc.
   b) Quick analytical monitoring of the river at strategic points.
   c) Predictions, through use of computer models, of the downstream concentrations of the spilled product at given times.
   d) Quick access to all health and safety information for products handled. In some cases additional data is being generated.
   e) Notification of concerned parties in the event of a spill, as well as later if water usage should be

suspended.

f) Work with the Coast Guard and others in handling the situation to minimize hazard to the public and the environment. This would include removal of pockets of product, where possible, from the top or bottom of the river utilizing equipment identified in advance as being available for this purpose.

Incidentally, we have already calculated that even if the largest load of the most toxic material shipped were completely lost into the water body at low flow and none were removed in the purification process, the concentration would be considerably below the point where lethality would be a problem.

Senator Muskie recently called for "one-armed scientists." It seems he was frustrated by witnesses from the National Academy of Sciences who said, "on the one hand, the evidence is so; but on the other hand..." I'm afraid I come to you as a two-armed scientist. I believe that industry is trying to solve problems. On the other hand, we recognize that there are still many problems to be solved. I will now speak of my current area of expertise, the health effects of chemicals.

For the vast body of chemicals, we don't have as much information on toxicology and medical effects, and environmental properties as we would like to have. Here is an example of how we are making the most of the present situation.

To deal with the problem of spills, we are developing some benchmark numbers as a guide to acute hazards in potable water. For example, allowable air concentrations have been defined for a large number of organic chemicals. These are the so-called "Threshold Limit Values" of the American Congress of Governmental Industrial Hygienists. As an example, the TLV of ethylenediamine is 25 mg per cubic meter. Since the material is highly soluble in water, we assume that 80% of the material would be absorbed in inhalation. Considering lung capacity factors and the body weights of man, we come up with a benchmark of 400 mg per liter of water as the ingestion equivalent of an inhalation TLV. In view of the fact that there is only minimal toxicity data on ethylenediamine, we scale this down to 4 mg per liter as the point where we could become very concerned about the quality of potable water resulting from a one-time spill. This calculation is based upon assumptions; but at least it is a reference point for scientists and engineers to cope with a spill.

I am sure that most of you are quite concerned about the fact that long-term chronic toxicity data does not exist for many of the chemicals in use today. In the U.S. industry, we are addressing this absence of data in two ways.

Individual chemical companies are increasing their toxicity testing in a rapid manner. Toxicological and environmental testing on chemical products increased nearly 100%

between 1972 and 1974 according to a recent study by Foster D. Snell. Similarly, a recent survey showed that ten companies will increase their expenditures for toxicological and environmental testing by 70% between 1974 and 1976.

Early this year, 17 chemical companies initiated a new organization, The Chemical Industry Institute of Toxicology. The Institute will address two important concerns: developing more extensive data on commodity chemicals, and developing new methodology in toxicology and in the diffuse area of translating animal results to man. The goal of this effort is to assist the total needed effort toward better evaluating occupational health and environmental effects of the building blocks of the chemical industry. The Institute, to be located at Research Triangle Park, North Carolina already has temporary committees of member company scientists actively formulating concepts for research.

SCIENCE ENGINEERING AND MANAGEMENT

Let's return to my theme of water problems and solutions in the context of history. The common thread of water problems seems to involve solutions that have been based upon sound applications of engineering and science along with flexible and varied value judgments. It seems to me that today's theorems for sound management must include:

A flexible system of reference values.

Predictive costing of various management control strategies.

Consideration of the new dimension of energy consumption on control strategies.

A scientific data-base for assessing health effects and aquatic effects.

Practical benchmarks of hazard in assessing health and environmental effects.

Common sense considerations of what's practical. For example, we withdraw large quantities of water from the Mississippi for use in cooling towers. Now one EPA proposal would require removal of the "natural" mud before the water can be returned to the Mississippi.

I am sure that each of you, with your varied speciality backgrounds, could lengthen this list with equally valid, practical considerations of water management.

VISION

This decade of the 1970's may well go down in history as one of the most dynamic ever -- at least in terms of value systems. We have already accepted the realization of practical and finite limitations to our energy resources. Whole new energy systems will be developed in the next 15 to 20 years. Just the construction of new power plants by yet undefined technologies will impose a negative energy utilization for several years. In this century, we have seen a multi-thousand ton organic chemical industry grow to a multi-million ton industry and in the process, convert from coal resources to petroleum resources. During the next 75 years, this industry will undoubtedly have to shift drastically, either back to coal or to agriculture for its raw material base. We may well approach the time when what we today call thermal pollution will become an asset for growing algae or shrimp or a means of irrigating crops and with the added benefit of accelerated growth because of the heat. Today, we have a few selected municipalities experimenting with disposal of municipal wastewater on the land -- but with three different motivations: to purify water; to provide crop nutrients; or to provide irrigation water for agriculture.

I have tried to illustrate in simple and common sense examples, that varied solutions must be found for our many, many water problems and challenges of the future. As leaders in the standard-setting process, I encourage you to keep your approaches to standards flexible; to standardize only what is really necessary for health and environment.

I encourage you to resolve areas of uncertainty by stimulating productive research and engineering. Don't prematurely get yourself locked into unreasonably small numbers without a clear understanding of natural background levels. For example, recent studies from the University of Hawaii have reported that edible seaweed contains significant quantities of bromoform and related materials. And similarly, foster a thorough understanding of the analytical methodology.

Each of you are positioned as responsible persons in the area of water quality. Current pressures certainly push toward standardization and, in my opinion, for overstandardization -- limited time regulations to <u>initiate</u> change rather than to shackle or tunnel change; standards which state the performance objective and leave the solution to the ingenuity of each wave of technology change and to competition.

It seems to me that setting standards today is largely a practice dictated by the expediency of in-hand solutions and the political weighting of legislative bodies and the press. Vision calls for standard-setting to foster long-term management ingenuity by both the public and the private sector.

Each of you will leave this symposium either with visionary challenges for dealing with water quality, <u>or</u> with the burden of standardizing toward stultification. I believe this symposium is providing valuable perspective for your choices.

SECOND DAY

OPENING REMARKS

DR. SMEETS: We shall begin this morning's session, and we have quite a lot of work to do today. We will have some very excellent papers that will be presented during this session. So, I think we must follow the time schedule. But before starting the technical part of this session, I should like to thank, on behalf of all the participants, Dr. Mrak and Dr. Coulston and Dr. Korte, for a very fine supper they offered us last night. I think we enjoyed the party very much, and we thank you much for the organization.

It is a great pleasure for me to invite Dr. Harold Egan of the Laboratory of the Government Chemist, to present his paper on "The Development of Analytical Standards for Water Quality." Dr. Egan is the Director of the Laboratory of the Government Chemist, which is one of the scientific research departments of the Department of Industry in England. Dr. Egan directs about 400 supporting scientific staff members, most of whom are analytical chemists.

It might be of interest to learn about the function of this laboratory, since I think this is a very special situation among our countries. The functions of this laboratory are first, to offer a comprehensive, analytical service to all government departments and, second, to give to all government departments direct scientific advice, based on analytical and other chemical considerations.

The government scientist has to carry out research in analytical chemistry where the development of analytic methods is of public importance. And the fourth point is to provide expert evidence in connection with the above and in accordance with the statutory analytical duties of the government chemist.

With all these functions, you see that we should be very happy to have the Director of this Laboratory in our midst today. We are very pleased to hear your paper, Dr. Egan.

THE DEVELOPMENT OF ANALYTICAL STANDARDS FOR WATER QUALITY

Harold Egan

INTRODUCTION

In considering the development of analytical standards for water quality, I will deal mainly with the contribution of chemistry and the analytical chemist. This is not intended to minimise the significance of other standards based, for example, on physical or microbiological considerations. Indeed the latter, which were perhaps the main area of endeavour in the development of water quality standards a century ago, are still essential to almost all use standards today.

To begin with, I would like briefly to look back at some of the early approaches to analytical standards in order to see how they have influenced our present attitude; and perhaps in particular to show how in recent years what have developed as traditional analytical criteria for water quality are now being supplemented by and in some cases are giving way to more modern such standards. I should at the outset take a general view of the position of analytical chemistry and the analyst in the assessment of the quality of the aquatic environment (1). Korte (2) has defined 'environmental quality' in relation to that part of the environment which is 'noticeable' in terms of amenability to description in physical or chemical terms. More recently a Working Group of the SCOPE committee of ICSU has defined it as 'the state of the environment as perceived objectively in terms of its components, or subjectively in terms of its attributes such as beauty and worth' (3). 'Environment' and 'health' are concepts which are closely related: and so also are environment and the general standard of living. Environmental quality and environmental health are cost-benefit situations: and like many such situations also raise political issues. The relation between chemical analysis and environmental quality is thus a complex and far-reaching subject.

The distinction between criteria and standards has been made elsewhere (35) and a conceptual framework for developing the latter from the former is available. One simple approach is to look at the use of chemical analysis as a basis for the realization of environmental quality criteria. This is not a matter for the analytical chemist alone; it calls for an understanding between him and the ecologist, the marine biologist, the plant pathologist, the human toxicologist and many others. However, the analyst can at least strive to provide basic indexes for many of the various criteria which need to be taken into account in evaluating environmental quality. In particular, he can measure the level of substances, usually trace substances, which intrude (or otherwise appear) in the aquatic or other environment; and what is often just as important, he can contribute to a sensible understanding of the trends (and the significance of these) for individual environmental factors and of the possible interaction between these factors.

DEVELOPMENT OF ANALYTICAL STANDARDS

The general introduction to the US Joint National Academy of Sciences National Academy of Engineering report "Water Quality Criteria 1972" (35) very briefly traces the historical background to the development of water quality criteria from prehistoric times. Whilst there had been analyses of waters for their mineral, dissolved gases and other contents previously, the systematic application of analytical methods to the quality assessment of drinking waters began to take shape during the first half of the nineteenth century and led, during the latter half of that century, to the conscious assembly of collections of such methods into textbook and reference book form. One of the earliest of such compilations was "Water Analysis" by J.A. Wanklyn, first published in 1868. Writing in the sixth edition of this book in 1884 (4), Dr. Wanklyn said that "for most sanitary purposes a water analysis is complete when it includes total solids, chlorine, free and albuminoid ammonia, oxygen consumed in moist combustion (a semi-quantitative alkaline permanganate distillation procedure) and poisonous metals". He gave detailed laboratory instructions for carrying out all of these processes and guidance as to the interpretation of the results. Analytical chemistry today has become perhaps the most important tool which, when coupled with toxicology, gives us the main scientific approach to safety aspects of environmental issues. Analytical chemistry is also a central instrument for the consideration of many of the amenity aspects of environmental problems - indeed there is no sharp dividing line between the two aspects of safety and amenity; scientific opinion and social judgment become merged. 'Pollution' and 'contamination' are relative terms, subjective in character. A pollutant has been defined as something that is present in the wrong place at the wrong time in the wrong amount. Place, time and amount are all involved. Analytical chemistry is widely used to provide indexes of these and is one of the disciplines which is in the centre of environmental issues at the present time.

I cannot today do more than review the subject but I have tried for the record to bring together some of the main analytical compilations and commentaries. Analytical methods for pollutants of the aquatic environment are of interest on a local scale as well as on a regional, national and possibly also a global basis. In Britain an official Standing Committee on Methods of Water Analysis has published an analytical manual (9) whilst in the United States guidelines establishing analytical procedures for the analysis of pollutants in water have been published in the Federal Register (10). Water quality is now the subject of consideration by the International Organization for Standardization's technical committee ISO/TC 147. The basic standards are those of the World Health Organization (7,8); these are set out in its main publications in relation to designated types of analytical approach.

QUALITY CRITERIA

The quality criteria and hence the standards for water will depend in part on the purpose which the water serves or is intended to serve, although considerations of multiple use frequently arise in practice. Drinking water, historically and not unnaturally, was the first use to receive regular and systematic attention. Surface water intended for the abstraction of drinking water, effluent and waste waters, irrigation water, water for animals, fresh water for bathing, amenity waters and indeed, marine waters are also the subject of quality standards. "Water Quality Criteria" (35) is a comprehensive volume offering guidelines for water quality standards in relation to recreational and aesthetic aspects, public water supplies, freshwater aquatic life and wildlife, marine aquatic life and wildlife, agricultural usage and industrial usage. Mancy (11) has listed parameters of general significance in water quality characterization, distinguishing between domestic water supplies, those for recreational and aesthetic purposes, those relevant for aquatic life, fish and wildlife and those for water for agricultural use. He distinguishes permissible and desirable quality levels for domestic supplies; and continuous and short-term use standards for agricultural purposes.

COMPILATION OF METHODS

There are many compendia of standard analytical methods for the assessment of water quality. I have already referred to the basic methods manuals in the United States (10) and Britain (9). Wilson (33) has prepared a substantial bibliography of official, standardized and recommended methods of analysis which covers most of the chemical indexes used. These methods are very often based on collaborative analytical studies, based for the most part on carefully designed system of sampling and replication of procedures within laboratories and between laboratories. There is a multiplicity of such studies at the present time and this in itself raises something of a problem. How does one decide between two or more analytical methods for the same thing when they have each been established by competent collaborative action! How does one compare such methods or is there now a need to harmonize the principles of collaborative studies? For some methods, the procedures need to be standardized exactly since the procedures themselves define the quantity being measured or assessed - BOD or TOC are examples. Other methods may best be approached by standardizing the analyst rather than the method, that is to say the choice of a method for such well defined chemical entities as cadmium, or DDT or chloride can be left to the professionally competent analyst. The standardization of the method (or the analyst) is not in itself enough, however. And whilst there is now perhaps a need to harmonize the philosophy of collaborative studies, there is also a need within a laboratory for quality control of analytical procedures to ensure that proper account is taken of such features as precision, sensitivity, selectivity and the repeatability and reproducibility of results.

An unusual compilation of methods is that published by the International Biological Programme includes practical details for methods which can be used in the field where only primitive laboratory facilities are available, for example in the extreme case of the limnologist arriving on foot at a lake with his laboratory in his rucksack; these are designated 'Level I'. Other methods are described for moderately well equipped laboratories with personnel who have a basic knowledge of analytical chemistry (Level II) whilst Level III methods include those which require more advanced equipment which is not at present routine in most laboratories (12). It is useful at this point to look briefly at some individual areas, fresh water, drinking water and rivers and oceans.

FRESH WATER

In some ways the examination of fresh water for environmental contaminants represents the simplest of analytical situations, since although the levels concerned may be minute the clean-up problems are, for fresh water at least, often minimal. This is in contrast, for example, to the examination of food or other biological material where the initial stage in the analysis is the removal of the great bulk of the organic material, a process which must be accomplished without loss of the traces of contaminant which it is desired to measure. Trace metals such as copper, lead and zinc can be measured spectrophotometrically on the water sample itself, directly or after only a simple concentration stage. Sometimes the concentration which occurs by natural processes: it simplifies the analysis or makes it more meaningful despite the clean-up stage involved. Thus persistent contaminants in water such as DDT may best be followed through indicator organisms which act as accumulators: aquatic organisms of interest in this connection include fish, molluscs and crustaceans.

The advantages and limitations of the various modern analytical techniques for trace metal pollutants - atomic absorption, atomic fluorescence, atomic emission and mass spectrometry, neutron activation analysis, X-ray fluorescence and anodic stripping voltammetry - have recently been reviewed by Coleman (13). Such methods can deal with quantities down to about one picogram per millilitre (in some cases even less, depending on the element) but of course they do not relate exclusively to the aquatic environment. Selective ion electrodes are also well established for many inorganic pollutants and the possibility of extending these to organic compounds by the use of immobilized enzyme systems has been described. An ammonium-selective electrode can, if covered with a layer of urease, become responsive to urea; other systems can be made to be responsive to glucose, lactates and aminoacids. These systems cannot, however, be more selective than the enzyme systems they involve. There are also many chromatographic systems, including the ring oven technique, for trace contaminants.

There are well-defined principles underlying the control of the discharge of radioactive wastes into the environment. Small volumes of high level radioactive wastes are stored in such a way that no leakage to the environment can occur whilst the activity of low level wastes is reduced, by dilution or decay, to a level at which they can be safely discharged (14). Natural radioactivity apart, three sources of radioactivity need to be considered in relation to environmental monitoring: effluent discharged to sea from nuclear fuel processing plants, cooling pond water (which may be discharged to sea or river) from nuclear power plants and radioactive fall-out resulting from nuclear weapon testing. Where extremely low levels of activity are encountered, precautions must be taken to reduce the background count rate to a minimum; these include the use of specially selected shielding materials having very low natural activity and of anti-coincidence counting systems to eliminate extraneous responses. In some ways low level radiochemical analyses are doubly difficult in that the preliminary clean-up process still relies on the older wet methods of chemical separation, which even with the aid of more modern techniques such as selective extraction still tend to make a lengthy, even cumbersome, process.

DRINKING WATER

As already mentioned, conventional analytical standards for drinking water have been elaborated by the World Health Organization, both on a European basis (7) and on a world-wide basis (8), the former being set at a somewhat higher level than the minimal values specified in the latter bearing in mind the extent of industrial development and intensive agriculture which have been reached in some European countries. These are both detailed advisory documents, which set out among other things tentative limits of toxic substances in drinking water and details of other substances and characteristics which affect the acceptability of water for domestic use. They do not give detailed directions for the methods of analysis to be used but there is a general indication of the methods by which the substances can be estimated and methods for the examination for physical, chemical and aesthetic characteristics of water. Limits of tolerances for piped supplies are also included in the European standards for toxic metals, together with details of chemical constituents which if present in excessive amounts (also indicated) may give rise to trouble. The World Health Organization is also preparing a comprehensive manual on analysis for water pollution control, to be published in 1977. The International Organization for Standardization (ISO) has established an environmental programme covering the areas in which international standards can make a valid contribution (15). ISO/TC 147 first met in Geneva in April 1972, when it was decided to restrict the scope of its activities to the definition of terms, and the sampling, measurement and reporting of water characteristics. For this purpose subcommittees and working groups were established to deal with particular aspects such as the physical, chemical, radiological, biological, microbiological and biochemical tests for water quality. The secretariat of ISO/TC 147 is held by the American National Standards Institute (ANSI). It is important to realize that whilst ISO is concerned with standardization in the field of water quality, it is not (unlike the World Health Organization, for example) concerned with setting actual standards of acceptability for water quality. Standards for surface water which are eventually intended for drinking have recently been proposed by the European Economic Community as the basis for a Directive (which is an instrument which is implemented by the individual member states by making or amending their own national laws). The proposals recognize three defined categories of surface water and set out standard methods of treatment for each, the frequency of sampling and the methods of analysis to be used being left to be defined by the national authorities (16).

RIVERS AND OCEANS

I have so far considered mainly the freshwater environment. The marine environment is much larger in extent and should also be considered. There are recent reviews of the environmental organic chemistry of oceans, fjords and anoxic basins (18), of hydrocarbons in the marine environment (19) and of the fate of DDT and PCB's in the marine environment (20). The organic chemistry of rivers and lakes has been reviewed recently by Cranwell (21). The first report of the Royal Commission on Environmental Pollution in Britain was concerned with river, tidal, estuarial and coastal waters (5), the latter aspect also being considered in some detail in the third report (6).

There are two broad aspects of chemical analysis in relation to the quality of the marine environment. The first I can designate simply as 'marine aspects' and relate to the analysis of the oceans themselves. The quality of the oceans has become of increasing concern in recent decades, in particular in regard to the presence of hydrocarbons (17), tar, oil, oil residues and, more recently, plastics.

The analytical methods for sea water are less well developed than those for fresh water and there are fewer comprehensive compilations of these. Freshwater analysis techniques may require modification: whilst freshwater samples can in some cases be examined direct, for example for trace metals such as copper, lead and zinc, by atomic absorption spectrophotometry, there are obviously some limitations in the case of sea water. Chelating agents may be used for these trace metals, with concentration by solvent extraction. In this way, trace inorganic contaminants can be estimated by atomic spectroscopy with a sensitivity of from 10 to 10^{-2} micrograms per ml for metals, depending upon the element; or, if graphite atomizer methods are used in place of flame ionization, from 10 to 10^{-2} (or even 10^{-3}) nanograms per ml. Limits of detection may be some ten times lower. Mass spectrometry can also be used down to about 1 nanogram per ml and gives a better coverage of elements.

Ion-selective electrodes are also available but must be used with some discretion: there is a temptation to apply this technique for "any old ion" but selectivity is a relative matter and electrode responses are sluggish in sea water.

Tar, oil and oil residues and plastics have also commanded a good deal of interest recently. Oil contamination is largely a surface problem and tends to be localized in character: the Institute of Petroleum in Britain has sponsored a useful book on the practical aspects of sampling, analysis and interpretation of oil pollutants and their individual characterization in the context of marine pollution (22).

Quantitative assessment of the tar and plastic waste distribution have been made in surface waters of the Pacific Ocean (23). Attempts have been made to estimate the amount of oil discharged (accidentally or otherwise) into the marine environment (24) but analytical methods for direct estimation have not been widely used, partly because oils may fractionate or degrade fairly rapidly and thus are not easily amenable to this approach. There is, however, a substantial international interest in analytical methods for the characterization of oils in pollution situations. The subject of polynuclear aromatic hydrocarbons in potable and waste waters has recently been reviewed by Harrison et al. (25), who pointed out that whilst it is far from certain that these compounds are of significance when present in trace amounts in drinking water, there is still a need for further research into both the levels which occur and their health effects.

The second aspect I can distinguish as 'maritime', as opposed to 'marine' aspects of analysis and concerns marine organisms. Whilst these call for consideration in their own right, they are also of value in the analytical context insofar as they may serve to concentrate contaminants by natural processes so that, as already described for freshwater, their analysis serves as a general pointer to marine environmental quality, even if there is no damage to the organisms themselves.

ANALYTICAL STANDARDS TODAY

The scope of analysis as a basis for quality standards for raw, potable and waste waters, developed from classical wet chemistry techniques is well demonstrated by the methods which are published in Britain by the Department of the Environment (9). This gives details of preliminary sampling details and methods of sensory assessment, followed by working descriptions of a wide range of general physical and chemical measurements. The latter include solid matter in solution and suspension and volatile and non-volatile matter on ignition, total solids, dissolved solids, total suspended solids and settleable solids. Other physical criteria include electrical conductivity, pH value, alkalinity and acidity and free carbon dioxide. Chemical criteria comprise total hardness and calcium, magnesium, sodium, potassium, aluminum, molybdate-reactive silica, chloride, fluoride, sulphate, orthophosphate, total inorganic phosphate, total phosphorus and residual chlorine, chloramines and nitrogen trichloride contents. General measures of organic pollution include dissolved oxygen, ammoniacal nitrogen, organic nitrogen, total unoxidized nitrogen, nitrite, total oxidized nitrogen and nitrite, anionic surface-active agents contents;

biochemical oxygen demand, chemical oxygen demand permanganate value and stability by methylene blue test.

Tests for particular inorganic pollutants include methods for traces of antimony, arsenic, barium soluble in dilute hydrochloric acid, boron, cadmium, chromium present as chromate, total chromium, cobalt, copper, ferric, ferrous and total iron, lead, manganese, mercury, nickel, selenium, silver, tin and zinc; and procedures for other selected pollutants including phenols, thiocyanates, cyanide in the presence of ferroxyanide or ferricyanide and total cyanide, formaldehyde, sulphide, sulphite, thiosulphate and pesticides; and for waste water volatile immiscible liquids and non-volatile matter extractable by light petroleum. In addition, a number of simple methods for small sewage works are described for testing sewage effluents. These include a field test for permanganate value, an approximate (4 hour) permanganate value, ammonia nitrogen by direct nesslerization, a qualitative test for oxidized nitrogen and the estimation of settleable solids by volume. Each of these many parameters of course affords some measure of water quality criteria in its own right; but there are also a number of compound or objective quality indexes combining two or more such measurements, as discussed for example by Harkins (12a).

Specific organic contaminants can call on a wide range of chromatographic techniques which, by their great versatility, offer the opportunity to approach virtually any trace environmental problem. The methods developed tend to fall into two groups, those responding down to about 0.1 mg per kg and those sensitive to 0.1 micrograms per kg. The persistent organochlorine pesticides such as DDT, for example, can be detected at levels down to about 1 microgram or less per litre. Such methods may be used not only for quality evaluation but as an aid to the understanding of transference mechanisms. It is, however, most important to realize that whilst it may be possible to detect traces at levels of 1 in 10^9 or even 1 in 10^{12}, there is also a need to establish the identity of the substances and this can be a very difficult problem. But it is also a very important matter in the context of long-term, low-level exposures. There are also well-established methods based on methylene blue reduction for the estimation of traces of anionic detergents in water. Two alternative procedures are discussed currently for non-ionic detergent residues, based respectively on quantitative thin layer chromatographic separation (26) and titrimetry (27).

There is still an interest in improving analytical techniques, particularly as regards accuracy and precision, as indicated in the United States report on research needs (36). Whilst the continuing knowledge of all or most of these characteristics is essential, an indeed at least one (nitrate) is becoming of increasing importance, there is today a growing and urgent requirement to judge potability not only on these specific constituents, but also on what I might call "packaged" analytical criteria. The measurement of total organic carbon (TOC) illustrates one facet of this trace work currently of interest. TOC is estimated by combustion to carbon dioxide (which is measured), a correction being made for any inorganic carbonate present if this has not previously been removed. TOC values are being introduced as an index of water quality, in addition to the older biological oxygen demand (BOD) and chemical oxygen demand (COD) values and the permanganate value (PV). BOD, COD and PV figures will probably continue to be used in order to maintain continuity of records, however; and whilst TOC values (which afford a true measure of the organic carbon content) can be measured rapidly to follow trends, there is doubt regarding their correlation with the older parameters. COD, BOD and PV tests are empirical and are time-consuming to perform and they may not take into account traces of persistent, non-biodegradable organic matter. There are variants of TOC. Total oxygen demand (TOD), for example, involves burning the sample with the aid of a catalyst at a temperature of about 900°C in a stream of carrier gas such as nitrogen containing a small and constant proportion of oxygen; and monitoring the residual oxygen. Organic carbon, nitrogen and hydrogen are thus oxidized with an efficiency greater than that achieved in the COD test; dissolved oxygen interferes and a correction has to be made for this (34).

In addition to the standard abstract journals such as Chemical Abstracts and Analytical Abstracts and the specialist water abstract journals such as Water Research Centre Information, critical reviews of methods of water analysis are published approximately every two years in the review issue of Analytical Chemistry. The most recent review, by Fishman and Erdman (28), covers the period from October 1972 through September 1974 and includes a note of other pertinent reviews of the subject, including the literature on water pollution control, published during this period. Also included are detailed reviews relating to analytical aspects of alkali metals; aluminium, iron, manganese and chromium; copper, zinc, lead, cadmium, nickel, cobalt and tin; mercury, silver, and gold; molybdenum, vanadium, bismuth, uranium, thorium and rare earths; boron, selenium, arsenic, antimony, phosphorus and silica; halides; sulphate and sulphide; nitrate, nitrite, ammonia, organic nitrogen and cyanide; oxygen and other gases; detergents and pesticides (including herbicides and fungicides); oxygen demand and total carbon and various individual classes of organic compounds including glycols, lignins, amino-acids and sugars, together with radioactivity and isotope analysis.

CONCLUDING REMARKS

At the international level, the International Council of Scientific Unions (ICSU) established a Scientific Committee on Water Research (COWAR) in 1964 to consider the problems of international water resources in all its aspects and to formulate programmes of research. The International Union of Pure and Applied Chemistry (IUPAC), a member of ICSU, has in its Applied Chemistry Division a Commission on Water Quality which is, among other things, concerned that chemical aspects of water quality are properly considered. IUPAC Committee on SCOPE (the ICSU Scientific Committee on Problems of the Environment) has recently published some selected analytical methods relating to water quality (29).

The pattern of work in the potable water field is now going through something of a state of transition, inevitable in view of the emphasis currently being placed on environmental pollution and on trace contaminants generally. The traditionally determined characteristics of a potable water such as pH, alkalinity, hardness, ammonia and nitrate, in addition to bacteriological examination and possibly one or two of the more common trace metals, have been universally catalogued as an indication of the suitability of a water supply for drinking for many decades; the traditional list probably originated in Wanklyn and the early editions of Tresh, Beale and Suckling's "Examination of Waters and Water Supplies", published at the beginning of the century. It may be of interest to compare the WHO recommended limit for lead in water of 0.1 mg per litre with that suggested by Wanklyn in 1884: "good drinking water should contain less than one-tenth of a grain of lead per gallon", equivalent to about 1.4 mg per litre.

Instrumentation for assessing water quality has developed with the expanding interest in the subject itself. While wet chemical analysis continues to be useful in many areas, including those which relate to research investigations or other calls for precise measurements such as instrument calibration, the need for rapid, frequent trace analyses has resulted in a great increase in the use of instrumental methods and in the development of automatic and continuous automated analytical processes of assessment of water quality. Instrumentation is also dealt with by the American Public Health Association in its standard methods book (30) and in the methods compilation published by the U.S. Environmental Protection Agency (31). Automated water analysis has been reviewed recently by Phillips, Mack and MacLeod (32), who classify instruments as continuous samplers, semicontinuous samplers and laboratory analyzers. Continuous monitors use sensors based on purely physical responses to detect a change in water quality parameters; semicontinuous analyzers use a measured portion of the water under examination, the constituent being tested being converted to a form suitable for measurement by one or more chemical reactions, the analysis being repeated on a regular basis. Mancy (11) had described *in situ* sensor systems and remote non-contact optical measurements for monitoring the water quality of rivers, lakes and effluents. There is an increasing need for new methods for new substances, or for existing substances the importance of which have taken on a new significance.

The number of determinations made to assess quality is growing rapidly as a result of the increasing use of water, the increasing production of effluents and an increasing interest in the aquatic environment as a whole. Analytical standards still play a very important role in the assessment of water quality. However, there is now a need for quicker methods, automatic methods and in some cases approximate methods which may not have the accuracy and precision of the more detailed procedures but which give results of sufficient accuracy for specific purposes.

REFERENCES

(1) H. Egan, *Chemistry & Industry*, 1975, 814.
(2) F. Korte, "Comparative Studies of Food and Environmental Contamination", 1974, 5, Vienna: IAEA.
(3) "Environmental Impact Assessment: Principles and Procedures", 1975, Paris: SCOPE.
(4) "Water Analysis" by J.A. Wanklyn, Sixth ed. Trubner & Co., London, 1884.
(5) Royal Commission on Environmental Pollution, First Report, Cmnd 4585, 1971, London: HMSO.
(6) Royal Commission on Environmental Pollution, Third Report, Cmnd 5054, London: HMSO.
(7) "European Standards for Drinking-Water", Second ed. World Health Organization, Geneva 1970.
(8) "International Standards for Drinking-Water", Third ed., World Health Organization, Geneva 1971.
(9) "Analysis of Raw, Potable and Waste Waters", Department of the Environment, HM Stationery Office, London, 1972.
(10) U.S. Federal Register 1973, *38*, 28758-28760.
(11) K.H. Mancy, "Analytical Problems in Water Pollution Control" in National Bureau of Standards Special Publication 351, pp. 297-382, Washington 1972.

(12) H.L. Golterman & R.S. Clymo (eds.) "Methods for Chemical Analysis of Fresh Waters", London, International Biological Programme.

(12a) R.D. Harkins, J. Water Pollut. Control Federation 1974, 46, 588; 1804.

(13) R.F. Coleman, Anal. Chem., 1974, 46, 989A.

(14) "Reports on Environmental Monitoring Associated with Discharges of Radioactive Waste", U.K. Atomic Energy Authority, London.

(15) "ISO and the Environment" International Organization for Standardization publication 72/00173, Paris 1972.

(16) EEC Council Director 75/44/EEC, O.J. No L194 (25 July 1975) pp. 26-31.

(17) J.W. Farrington & P.A. Meyers, Environmental Chemistry, 1975, 1, 109.

(18) R.J. Morris and F. Culkin, Environ. Chem., 1975, 1, 81-108.

(19) J.W. Farrington and P.A. Meyers, Environ. Chem., 1975, 1, 109-136.

(20) M.M. Rhead, Environ. Chem., 1975, 1, 137-158.

(21) P.A. Cranwell, Environ. Chem., 1975, 1, 22-54.

(22) Institute of Petroleum, "Marine Pollution by Oil", 1974, Barking: Applied Science Publishers.

(23) C. Swong, D.R. Green and W.J. Cretney, Nature, 1974, 247, 30.

(24) R.M. Atlar and R. Bartha, Residue Reviews, 1973, 49, 1.

(25) R.M. Harrison, R. Perry and R.A. Wellings, Water Research, 1975, 9, 331-346.

(26) S.J. Patterson, E.C. Hunt and K.B.E. Tucker, J. Proc. Inst. Sewage Purif., 1966, 2, 190.

(27) R. Wickbold, Tenside, 1972, 9, 173.

(28) M.J. Fishman and D.E. Erdmann, Anal. Chem., 1975, 47, (5), 334R-361R.

(29) "Environmental Pollutants: Selected Analytical Methods" SCOPE, Paris, 1975.

(30) "Standard Methods for the Examination of Water and Wastewater" 13th Edition. American Public Health Association, Washington, D.C., 1971.

(31) "Methods for the Chemical Analysis of Waters and Wastes", U.S Environmental Protection Agency, Cincinnati, Ohio, 1971.

(32) S.L. Phillips, D.A. Mack and W.D. MacLeod, Anal. Chem., 1974, 46, (3), 345A-356A.

(33) A.L. Wilson in "Official, Standardized and Recommended Methods of Analysis", pp. 847-854. Society for Analytical Chemistry, London, 1973.

(34) B.D. Ravenscroft, Proc. Soc. Anal. Chem., 1975, 12, 254.

(35) "Water Quality Criteria 1972", Environmental Protection Agency, Washington, D.C. (1972).
(36) "Research Needs in Water Quality Criteria 1972", National Academy of Sciences, Washington, D.C. (1973).

INORGANIC TRACE ELEMENTS AS WATER POLLUTANTS;

Their Implications to the Health of Man and the Aquatic Biota

J.K. Miettinen

I. Inorganic trace elements and man's health

Man's activities are measurably changing the chemical make-up of air and water. These changes are increasing with the growing world population and the per capita consumption of energy and materials. Everything that man injects into the biosphere - inorganic, organic or biological - may finally reach natural waters. Inorganic trace elements are of special concern because they are nondegradable and, therefore, persistent. Some of them, particularly some of the heavy metals, are highly toxic and can affect even in low concentrations the aquatic life and the food-webs depending on it, the usability of water for irrigation purposes, for inductrial raw water, or as drinking water for livestock and man. It is the task of the environmental protection agencies to check this sequence of pollution effects at different levels. Drinking water is the very last checking point, but for man's health, an important one because in technologically developed countries, 3/4 of the population consume tap water today. The earliest and best would be, of course, checking at the source.

Grouping of inorganic elements on the basis of their effects on man's health

There are 83 stable elements and a few natural very long half-life radioactive elements, particularly U and Th, on the earth. The stable elements we can broadly divide into three groups:

1. The essential elements - three macro elements, Ca, P, Mg and, fourteen trace elements (1).
2. The toxic elements - in practice, about 10-15, for five of which there are international standards in drinking water (As, Cd, Pb, Hg, Se) (2) (of these, selenium is also in the list of essentials), and, when the five stable noble gases and the four light elements H, O, C and N, are excluded.
3. The non-essential inorganic elements - numbering about fifty, many of which are also toxic, in some instances. By heavy metals, was originally meant metals heavier than vanadium (Z=23), but nowadays, often all elements heavier than it, sometimes even such light toxic metals as beryllium and such metalloids as arsenic, are included to them (see pt. II).

4. The natural radioactive elements - and man-made radioisotopes, can be conveniently classified as the fourth group of inorganic trace elements.

I begin my review by briefly characterizing these groups and the role of the elements on man's health.

Essential trace elements

Presently, fourteen trace elements are believed to be essential for animal life: Fe, I, Cu, Zn, Mn, Co, Mo, Se, Cr, Ni, Sn, Si, F and V (1). A recent report by the World Health Organization (WHO) (1) states that "Classification of the trace elements into essential, non-essential and toxic groups can be inaccurate and misleading. All the essential elements become toxic at sufficiently high intakes, and the margin between levels that are beneficial and those that are harmful may be small... It would not be surprising, therefore, if other trace elements classically regarded as toxic elements were also found to be beneficial or essential."

Of the above fourteen elements, the last five (Ni, Sn, Si, F and V), were added in the last six years, following the introduction of better purified experimental diets.

Man's needs for trace elements are not yet precisely known. Of the above list of essentials for animal life, silicon has not yet been shown to be important for man. There is increasing evidence that lithium and boron, too, may play a role in animal nutrition. The Food and Nutrition Board of the US National Academy of Sciences - National Research Council, has recently provided a new edition of its list of Recommended Dietary Allowances (RDA) for seven minerals, three macro- (Ca, P, Mg) and four micronutrients (I, Fe, Cu, Zn), for which the nutritional requirements are most precisely known (Table 1). As can be seen, these requirements vary considerably for different physiological states. Iodine and fluorine are two essential elements the deficiencies of which were shown to be associated with the epidemic diseases goitre and dental caries, respectively. Both relationships were elucidated by geographic studies of the mineral availability and the frequency of the associated disease, and both diseases can be corrected by adding the missing nutrient to food or drinking water. Information on the requirements, as well as the noxious or toxic levels of intake, would be necessary for all elements in order to gain knowledge of the range in which the intake may or should vary, but as yet, such information is lacking, even for most of the essential trace elements.

The toxic elements

It was mentioned earlier that there are international standards for five toxic elements in drinking water (Table 2).

In addition, international limits have been drawn for six essential metals (Ca, Cu, Fe, Mg, Mn, Zn) which may affect the acceptability of water for domestic use (Table 3) (2).

Zinc and copper in tap water often come from new galvinized piping, and lead from old piping. According to Rössner (3), toxic concentrations of Cu are not possible in drinking water, but it has to be limited because of its unpleasant taste. The earlier reports that copper is toxic were probably due to As or Pb (4).

Chloride, fluoride, sulphate and total hardness are also regulated by the WHO (2). For fluorides, the lower and upper limits vary from 0.9-1.7 mg/l, respectively, at air temperature 10-12°C to 0.6-0.8 mg/l respectively, at 26.2-32.6°C.

In addition to the toxic substances in Table 2, WHO recommends that other toxic metals such as Ba, Be, Co, Mo, Sn, U and V should be controlled in drinking water, though insufficient information is at present available to enable tentative limits to be given for these elements (2).

The non-essential elements

Less is known about the significance in water of the elements of the third group - about fifty stable elements which were not classified as "essential" or "toxic". Little is known about their metabolism and significance in animals and man. A recently published ICRP report, No. 23 (5), summarizes balance data on 51 elements, most of which belong to this group.

Health and minerals in water

During the last seventeen years, great interest has been shown in the relationship between health and mineral characteristics of local water supplies. This interest originated from the studies of Kobayashi in Japan (6) and Schroeder in the USA (7,8). In Japan, mortality from cerebral hemorrhage was directly related to the acidity of river water in the area. In the USA, water softness showed positive correlations for cardiovascular disease and arteriosclerotic heart disease. Similar correlations were later found in Great Britain (9), Sweden (10), the Netherlands (11), and Canada (12), but some studies did not corroborate such a correlation. The extensive and somewhat contradictory literature has been reviewed by several authors (eg. 13, 14, 15).

There seems to be considerable evidence suggesting a negative correlation between hardness of water and mortality from cardiovascular disease. Hard water may inhibit, but soft water promotes extraction of harmful elements from soil or distribution pipes. However, hardness seems not to be the complete answer. Schroeder (15) suggests that "Langlier's

Index" for corrosiveness of water be substituted for hardness and further studies directed towards discovering what cardiotoxic metals are dissolved by soft water from pipes.

In addition to high concentrations of Ca and Mg, which usually are the cause of hardness, such trace elements as Cr, Mn, V and Zn are often labelled beneficial, while Pb, Cd, and Cu are considered to be harmful. The complicated relationships between these elements are illustrated by Neri *et al.* (16) in Figure 1.

Lithium is not interrelated with the other trace elements but has been found beneficial in some American studies. It is believed to reduce the stress factor and in this way, the risk of myocardial infarction.

Cobalt and selenium decrease myocardial magnesium and copper. Selenium protects against toxic effects of mercury and cadmium. Calcium decreases the absorption of lead and cadmium. Copper deficiency decreases tolerance to cadmium, while excees of copper can induce zinc-deficiency. An increased Zn:Cu ratio has been found experimentally to produce an increase in serum cholesterol, while a decrease of the Zn:Cd ratio may increase blood pressure (for reference, see 16).

Many trace elements are important in biological processes. They can activate enzymes, compete with other elements for binding sites, influence the permeability of cell membranes, etc. Evidently, subtle deficiencies or excesses may lead to suboptimal health and subclinical, severe or even fatal chronic diseases. The causal relationships and mechanisms are still very incompletely known. Intake takes place often via several routes - inhalation, drinking and food. The total intake is decisive, of course. If it is excessive, the first thing to know ought to be the significance of each route of intake. Only then can corrective measures be designed in the most effective way.

The ecological correlations of the type described above have been useful by drawing attention to some elements of suspicion, but they have little further to contribute. They alone are not likely to reveal any specific causal relationships. First, the numerous intercorrelations (eg. lithium, sodium, calcium, magnesium, etc., with hardness/softness etc.) mask the possible specific correlations. Second, the statistical data for elements in air, water, foods and soil are usually obtained by different agencies or faculties, and often provide a poor basis for computing the intake of populations, not to speak of individuals. For instance, the concentrations of copper, zinc and cadmium in the water from the kitchen tap may be twice as great as in the source of water (17). Urbanized Europeans (and probably Americans, too!) consume only part of their total water intake as local tap water (18) (see Table 4). This percentage varies greatly, according to individual habits, as does the quality of the other drinks (milk, beer, wine, mineral waters, etc.). Bottled mineral waters may sometimes contain high levels of such inorganics as lithium, beryllium, arsenic, etc. (18). And finally, there are many other risk factors for C.H.D. besides the possible trace mineral imbalance, such as age, inherited predisposition, smoking, obesity, inactivity, etc.; it is not always possible to elminate the influence of all these factors, except in a study in which the group is composed on the basis of individual study.

Better results can be expected when populations from areas characterized by high and low mortality from coronary heart disease (C.H.D.) are analyzed by such an individual study. Such a study was recently performed by Pusnar *et al.* (19) in the eastern and western Finland. In eastern Finland, the mortality caused by C.H.D. is high and the soil, largely morainic, low in many trace elements. In both areas, households draw water from their own wells. The water, as consumed by the individuals, was analyzed. A cohort study on two rural male populations from these two areas, showed that in the period 1959-69 the death rate from the C.H.D. was in the eastern cohort more than twice as high as in the western cohort. The difference was connected with the quality of well water. Lower concentrations of chomium and higher concentrations of copper in the eastern area correlated most conspicuously with the high death rate from C.H.D. The concentration of serum cholesterol also correlated negatively with the chromium in drinking water in the eastern area. It has been recently shown, too, that the concentration of magnesium in the soil is much lower in the eastern (<100 mg/l soil) area than in the western area (>300 mg/l soil) (20). A similar multifaculty-multielement study of C.H.D. was organized in 1969 by a Joint WHO/IAEA Research Project in Relation to Cardiovascular Diseases. Its early results have been recently reviewed (21).

I have described to some extent the attempts to elucidate the effects of inorganic substances on man's health at the last check-point: drinking water. I now turn to the field in which our laboratory has been working, the effects of heavy metals on the aquatic life and the effect of one metal on the toxicity of another.

II. Heavy metals and aquatic life

The rapidly increasing consumption of energy and materials and the concentration of this consumption in heavily industrialized and urbanized areas has often resulted in severe pollution of the surface and even ground waters of the area.

An excess of organic materials, which depletes the oxygen,

and an excess of plant nutrients, particularly phosphorus, which cause eutrophication, soon become easily visible. Less conspicuous is the creeping pollution caused by inorganic trace elements. Of these, the group of "toxic heavy metals" has turned out to be most important in practice.

Originally, heavy metals were those having a specific gravity greater than 5 g/cm^3 (22). Later, other toxic elements, some of which were not even metals, were put under this heading. Zemansky (23) included in heavy metals all elements with $Z>23$, except the alkalis Rb, Cs and Fr, the alkaline earths Sr and Ba, and Y. Sometimes even lighter toxic metals like beryllium are included in "heavy metals"; here it would be better to speak of "toxic trace elements", however.

Mercury, lead and cadmium are usually considered to be the most dangerous ones as water pollutants when all factors are considered. I shall, therefore, limit my review primarily to them. Vanadium, indium, nickel, chromium, beryllium and silver are also highly toxic to the aquatic life. In the waste water of metalworks, the heavy metals have been put in the following order of decreasing risk, all factors (use, solubility, toxicity, etc.) considered: $Cd>Ni>Cr>Cu>Zn$ (24). Many other elements are highly toxic, but fortunately, most of them are so rare (and expensive), for instance, the noble metals and the lanthanides, that they are unlikely to exist in waters in harmful concentrations.

In unpolluted surface waters, heavy metals are normally present in very low concentrations. The calculated global averages for river water are presented in Table VI. For cadmium values less than 1 μg/1 (26).

Even in rainwater, their concentrations are not much higher, except in regions with heavy air pollution. An exception is lead. All rainwater is presently polluted by lead because of the use of leaded gasoline. Murozumi _et al_. (27) measured even in Greenland 0.2 μg/1 in 1965, a level about 400 times higher than in glacial ice from the year 800 B.C. In polluted rivers, lakes and coastal marine waters, heavy metal concentrations can be several orders of magnitude higher than in remote, unpolluted waters.

Mercury has proved to be the most dangerous aquatic pollutant because it is always rapidly methylated in the aquatic environment and in this form, it becomes accumulated in the food chains. The "enrichment factor" can be of the order of hundreds of thousands. There seldom exists enough soluble mercury in surface waters - except near a source - to disturb the use of the water for even drinking purposes, if it is otherwise clean or purifiable. Methylmercury can cause serious poisoning of food chains. While natural concentrations exceeding 1 mg Hg/kg fr.wt. can only be found in such long-living fish as swordfish and tuna and natural concentrations in lake fish seldom exceed 0.5 mg/kg, in Minamata, values up to 20 mg/kg of fresh weight were measured in fish and shellfish. In the USA and Canada, maximum values of 10-12 mg/kg, in Sweden and Norway 8-9 mg/kg, and in Finland, about 6 mg/kg have been reported from polluted waters. In Finland, the above value was measured in 1968. In the same year, the cellulose industry discontinued the use of phenylmercury as a preservative of the wet pulp and in the years 1970-72, the aquatic releases by the chlorine industry were drastically reduced. For instance, aqueous releases to the Kymi river were reduced from 670 kg/y in 1967 to 11 kg/y in 1970.

Since then, discharge to the atmosphere have also been reduced to some extent so that the total annual release is of the order of one per cent of the natural fallout in rainwater in Finland. At the end of the 1960's, the average level in pike in the most polluted area in Finland, Ahvenkoski, at the mouth of the river Kymi (near Kotka), was 3 mg/kg fr.wt.

In summer 1975, the level had decreased to 0.7 mg/kg. At the mouths of the Oulu and Kokemäenjoki rivers, the level has also decreased during the 1970's to about half, but in the lakes, the decrease is somewhat slower (Häsänen, 28). Rapidly changing water and fast sedimentation decrease the levels in river mouths faster than in lakes. The half-times observed in these waters, 3-7 years, are much shorter than were believed earlier.

The elucidation of the mercury risk in Scandinavia in the middle of the 1960's, thanks to the knowledge of the earlier case of Minamata, and its successful elimination by preventing the release at its source, dramatically demonstrate how worthwhile the measures for environmental protection can be. No observable health hazards to fish consumers occurred, thanks to the preventive action. Great economic hardship was borne by many innocent fishermen, however, and unfortunately, society has proved unable to compensate them properly, so far.

Cadmium is a relatively rare heavy metal (26). In the earth's crust, the concentration of the metallurgically important heavy metals decreases in the order Zn>Cr>Cu>Ni>Pb>Co>Cd. Natural waters contain only low concentrations of cadmium, usually less than 1 µg/l, although values above 10 µg/l have been reported both in natural and tap water, in the latter case, usually originating from the distribution pipes (26). In rivers polluted by cadmium, the metal may be undetectable in filtered water, particularly if the pH is neutral or alkaline, although the particulate fraction contains it in high concentration (29). The itai-itai disease on the Jintsu river in Japan was probably due to the transport of cadmium-containing suspended particles to the paddy soil by irrigation with river water (29). Shellfish can accumulate cadmium from bottom sediment but fish do not accumulate it efficiently (30). However, in chronic pollution, there is a risk of accumulation because a small percentage (in fish, about 1%) of the intake is retained very persistently (32). Thus, Lucas et al. (33) report in a variety of fish in the Great Lakes 0.06-0.14 mg/kg and Hartung, in the lower Mississippi, 0.024 mg/kg (34). In man, the biological half-time of the persistent fraction seems to be of the order of 15-20 years (31). The first symptoms of chronic cadmium poisoning occur in the tubular system of kidney. The critical level is about 200 mg/kg (26). The normal value for the present population is about 10-20 mg/kg (26). The bulk of the intake is from food and amounts to about 50-60 μg Cd/day (26). Inhalation normally adds only about 0.2 μg/day. Drinking water adds normally very little, but polluted water containing around 20 μg/l would increase the daily intake significantly, by 20 to 40 μg/day. Smoking 20 cigarettes per day will cause the inhalation of 2 to 4 μg of cadmium. This element is typical of those which can have many routes of intake. The liver of fish or shellfish from polluted water and of horse, elk, reindeer, cattle or sheep feeding on polluted forage may contain up to 200 mg Cd/kg fr.wt. (26,30). One meal of such liver would give more than one year's normal diet!

In 1968, the world production of cadmium was 14000 tons per year, and the increase is about 10 percent per year. Thus its annual production should be, in 1975, about 30000 tons. In 1970, the atmospheric emissions were 1/3 of the production. Half of the emissions originated from production of cadmium, zinc, lead and copper. Since then, the emissions have been reduced. The burning of coal and oil and, particularly, of plastic wastes, injects significant amounts of cadmium vapours into the atmosphere. In industrialized regions of Japan, cadmium fallout is several mgs per m^2 per month; in North Finland it is 1 $\mu g/m^2$/month. Thus atmospheric pollution can add significantly to the pollution of surface waters. Already, 0.01 mg Cd/l retards the growth of aquatic plants, and 0.1 mg/l is lethal (35). The same concentration is lethal to oysters. Cadmium in waters, rainfall, the aquatic fauna and the total diet has to be carefully checked. Its rapidly increasing use, very difficult recovery, long half-time in man, and relatively narrow safety margin in the populations of industrial countries demand extremely careful control.

Lead: Presently, the technological sources inject lead into the atmosphere at about 100 times the natural rate (36). Atmospheric fallout is an important mode of addition of lead to surface waters. Traffic, smelters and incinerators are the three main sources of atmospheric lead pollution. Lead concentrations in natural waters are presented in Table 6. Rivers running through industrial regions can contain as much as 120 μg Pb/l (41). The Rhine brings into the North Sea annually 2000 tons of lead, while the estimated atmospheric fallout into the North Sea is 1000 tons/year (42). The bulk of lead in surface waters is particle-bound. It is relatively rapidly sedimented. The sediments of the Great Lakes contain 1-200 mg Pb/kg (43). The lead concentration in sediment is usually 3 or 4 orders of magnitude higher than in water.

Fortunately, lead is rather poorly accumulated into aquatic food chains. The concentrations decrease along the food chain. Zooplankton can contain up to 1300 mg Pb/kg (44), plankton feeders, much less. Fish consuming bottom animals contain more lead than fish of prey. In the Great Lakes of Canada, fish meat contained 0.02-0.46, fish liver, 0.05-3.2 mg Pb/kg (45).

Lead is toxic for aquatic organisms, but less so than cadmium and mercury (Ravera, 46). It is usually particle-bound and does not inhibit the production of phyto-plankton. Fish avoid water containing 0.3 mg Pb/l (47) and above 1 mg/l is lethal (46). Such concentrations are usually possible only locally. Aquatic animals can hardly accumulate so much lead that their meat would become dangerous to eat, except some shellfish. If lead pipes are used, tap water may exceed the tolerance limit 0.1 mg/l. More common than food-poisoning, these days, are occupational risks and, for small children, risk from pica, eating of old paint. The dietary intake in industrialized countries is currently 200-550 μg/day (Table 7, 48), 10 to 20 times the natural level.

Lead may be an essential element to animals. It is chemically versatile, rather abundant, and not very toxic. It is essentially difficult to prove because the need can be small and the whole biota is so polluted by lead that it is very difficult to compose a leadfree but otherwise complete diet. The use of lead in water pipes, pewter, ceramic glazers, paints, etc. has put the mankind under stress from lead for over 2000 years. Many lead poisoning epidemics are known from history. The subclinical effects are a question of debate even today.

I consider lead the most dangerous heavy metal pollutant to mankind, but mainly as an air pollutant. Of the three main sources, smelters, incinerators and leaded gasoline, the first two should be rapidly stopped - if necessary, by closing those which cannot be provided with filters. Lead in gasoline should be reduced at a pace which is technologically possible. The aquatic releases should be controlled, considering the cost/benefit ratio. Lead is no enriched in the aquatic food chains.

Effect of metals on the biological activity of other metals

One metal may affect the effects of other metals. Potentiation is called synergism, weakening of the effect of another agent is called antagonism. Environmental pollution is usually caused by several metals simultaneously, and even by other pollutants. The need for "synergism studies" has been evident for some time, and few have been reported.

Hutchinson (35) observed that zinc increases cadmium's toxicity and accumulation in aquatic plants. Other authors observed zinc to reduce the toxic effect of cadmium in *Aspergillus niger* (49). Copper was found to increase cadmium toxicity (50).

Eisler and Gardner (51) by studying mortality of mummichog *Fundulus heteroclitus* by zinc, copper and cadmium in various mixtures, obtained results which are difficult to interpret, but which they describe as synergism between Cu and Zn and between Cd and Cu or Zn or both. The highest mortality with cadmium was obtained when the Zn:Cu ratio was 1 or 2. The highest cadmium concentration is surviving fish was found with low Cd concentrations in water at a Zn:Cu ratio of 36, and with a high Cd-concentration in water at a Zn:Cu ratio of 0.8. Copper-zinc synergism was also observed in rainbow trout (*Salmo gairdneri*) by Lloyd (52). Sprague (53) reviewed in three papers, toxicity tests on fish with various pollutants, including Zn, Cu and Cd. Although four-day lethality tests are common, 7-14 days periods may be preferable for these metals. Studies of chronic synergism are very difficult and very rare.

In our laboratory, several studies in this field have been carried out in the last few years. Pietiläinen (54) studied by laboratory tests the acute effect of lead and cadmium salts on the basic production of phyto-plankton and heterotrophic bacteria in the brackish water of the Gulf of Finland (salinity 0.55%). The basic production in the clean water, measured by the ^{14}C-method, varied between 1400-4000 mg C per m^3/day, according to time and place. All samples represented "strong" production. Blue algae represented 60-90% of the phyto-plankton.

Up to 0.3 mg/l lead had no effect (although 0.1 mg/l stimulated in one experiment), but 1 mg/l reduced by 50% and 10 mg/l by 90% the basic production. Even 0.1 mg Cd/l reduced the basic production to about 65%, 10 mg/l to 25% of the control. Both lead and cadmium increased slightly the dark fixation at low concentrations. The synergism was studied by keeping the concentration of one metal constant at 0.1 mg/l and varying the other from 0 to 3 mg/l. Synergism was observed when Pb:Cd ratio was $\leqq 3$, antagonism when it was $\geqq 10$. In both cases, the effect was small, about 20%. At a low Pb:Cd ratio (= high cadmium concentration) bacterial growth was greatly increased. Production in darkness was increased 18-fold when compared with the production in the presence of one of the metals only. No such increase was observed when there was more lead than cadmium in water.

Synergism of Zn+Hg and Zn+Cd in rainbow trout (Salmo gairdneri) has been studied by giving a sublethal chronic load of zinc to the fish by uptake from water. 1.58 mg Zn/l water is lethal within 6 days, but 1.34 mg/l is not lethal at all. The fish have lived in this concentration several hundred days without any other symptoms than slower growth than in the controls. In 40 days at such concentration, Zn contents of 10 mg/kg in bones, 20-30 mg/kg in blood, kidney, spleen, liver and gills and 60 mg/kg in skin are reached. After such an incubation, the Zn-containing fish have been submitted to various concentrations of radiolabelled mercury or cadmium (administered by catheter into the stomach) and its lethality compared with that in normal (low-zinc) fish. Accumulation of the labelled metal was followed by whole-body counting of the live fish. By equal applications, more cadmium was retained in the zinc-containing fish (55). By several administrations (0.1 ml/10 g fish) of a solution containing 2.21 mg Cd/ml, the normal low-zinc fish died in 62±3 hrs, the zinc-containing fish, after about 360 hrs. After death, the former fish contained 6.8 mg Cd/kg, the latter 22.9 mg/kg. Thus the zinc had a very remarkable protective effect (55). In a similar study with mercury, some (although less notable) protection by zinc, but mostly increased mercury uptake, was observed (56). Similar protection by zinc against cadmium toxicity is known. Parizek observed it in rats (57). It can be interpreted by assuming that zinc induces a high concentration of metallothioneine in the animals, and the other metal (Hg or Cd) becomes toxic by replacing part of the zinc from this metalloprotein.

Such synergism experiments, as briefly described here, are very difficult to design so that they would give clear results. The experiments should be of simple design and provide proper controls. The lower the concentrations studied and the more subtle the effects, the more difficult are they to identify. This is the dilemma of environmental poisoning, which causes subtle, marginal effects, but eventually, in large populations.

What are the priorities?

Environmental pollution, particularly elucidation of the subtle, chronic effects of the numerous trace pollutants, is a challenge to society. Heavy metals are usually close to the top of the lists of priorities.

What type of research would be most useful? We have already noted that epidemiological studies based on human mortality statistics cannot identify specific causes, although they may suggest what should be looked at.

Perhaps the most important task would be to gain more basic knowledge on the interplay of trace elements at the cellular level. Identification of the enzyme systems involved by simple experiments with cheap, easily-maintained organisms might be the most fruitful approach.

Some criteria generally applicable to the allocation of priorities in research on environmental pollutants - this author's choice - are presented in Table 8. The Nato Conference on Eco-Toxicity of Heavy Metals defined the priorities according to Table 9. The most important is problem oriented research strategy which utilizes the capabilities of various sciences, but is specifically designed for the task. Analytical chemistry is no longer a bottleneck. Instrumental, automatic, multielement analysis can clear a tremendous number of samples in a short time with a satisfactory sensitivity. Some comparisons are presented in Tables 10 and 11. Comparing sensitivities of different methods is a sensitive issue itself. The tabulations are often quite theoretical if the matrix is not the same and the sensitivity reported is one obtained from real samples. This is why some of the sensitivities of eg. nuclear activation analysis in Table 11 are much lower than the highest values reported in the literature. Most important is the representativeness of the samples. This is why the stress has to be in problem-oriented research strategy.

Acknowledgement: The research reported from the author's laboratory was financed by the IAEA, Contract No. 116/RB, and by the Academy of Finland, which is hereby gratefully acknowledged.

TABLE 1. Recommended Dietary Allowances (Minerals)[1] Designed for the Maintenance of Good Nutrition of Practically All Healthy People in the USA

	Years	Weight (kg)	Weight (lb)	Height (cm)	Height (in)	Energy (kcal)[2]	Minerals: Calcium (mg)	Phosphorus (mg)	Iodine (μg)	Iron (mg)	Magnesium (mg)	Zinc (mg)
Infants	0.0-0.5	6	14	60	24	kg x 117	360	240	35	10	60	3
	0.5-1.0	9	20	71	28	kg x 108	540	400	45	15	70	5
Children	1-3	13	28	86	34	1300	800	800	60	15	150	10
	4-6	20	44	110	44	1800	800	800	80	10	200	10
	7-10	30	66	135	54	2400	800	800	110	10	250	10
Males	11-14	44	97	158	63	2800	1200	1200	130	18	350	15
	15-18	01	134	172	69	3000	1200	1200	150	18	400	15
	19-22	67	147	172	69	3000	800	800	140	10	350	15
	23-50	70	154	172	69	2700	800	800	130	10	350	15
	51+	70	154	172	69	2400	800	800	110	10	350	15
Females	11-14	44	97	155	62	2400	1200	1200	115	18	300	15
	15-18	54	119	162	65	2100	1200	1200	115	18	300	15
	19-22	58	128	162	65	2100	800	800	100	18	300	15
	23-50	58	128	162	65	2000	800	800	100	18	300	15
	51+	58	128	162	65	1800	800	800	80	10	300	15
Pregnant						+300	1200	1200	125	18+8	450	20
Lactating						+500	1200	1200	150	16	450	25

[1]The allowances are intended to provide for individual variations among most normal persons as they live in the United States under usual environmental stresses. Diets should be based on a variety of common foods in order to provide other nutrients for which human requirements have been less well defined. See text for more detailed discussion of allowances and of nutrients not tabulated. From Food and Nutrition Board, National Academy of Sciences-National Research Council. Revised 1974.

[2]Kilojoules (kJ) = 4.2 x kcal.

Table 2. TENTATIVE LIMITS FOR TOXIC SUBSTANCES IN DRINKING-WATER. WHO, 1971.

Substance	Upper limit of concentration, mg/l
Arsenic (as As)	0.05
Cadmium (as Cd)	0.01
Lead (as Pb)	0.1
Mercury (total as Hg)	0.001
Selenium (as Se)	0.01

Table 3. SUBSTANCES AND CHARACTERISTICS AFFECTING THE ACCEPTABILITY OF WATER FOR DOMESTIC USE

Substance or characteristic	Undesirable effect that may be produced	Highest desirable level	Maximum permissible level
Calcium (as Ca)	Excessive scale formation.	75 mg/l	200 mg/l
Chloride (as Cl)	Taste; corrosion in hot-water systems.	200 mg/l	600 mg/l
Copper (as Cu)	Astringent taste; discoloration and corrosion of pipes, fittings and utensils.	0.05 mg/l	1.5 mg/l
Iron (total as Fe)	Taste; discoloration; deposits and growth of iron bacteria; turbidity.	0.1 mg/l	1.0 mg/l
Magnesium (as Mg)	Hardness; taste; gastrointestinal irritation in the presence of sulfate.	30-150 mg/l x)	150 mg/l
Manganese (as Mn)	Taste; discoloration; deposits in pipes; turbidity.	0.05 mg/l	0.5 mg/l
Sulfate (as SO_4)	Gastrointestinal irritation when magnesium or sodium are present.	200 mg/l	400 mg/l
Zinc (as Zn)	Astringent taste; opalescence and sand-like deposits	5.0 mg/l	15 mg/l

x) Not more than 30 mg/l if there are 250 mg/l of sulfate; if there is less sulfate, magnesium up to 150 mg/l may be allowed.

Table 4. The role of local tap water, non-local drinks and non-drinking water for urbanized Europeans. Zoeteman and Brinkman, Netherlands (18).

Route	Mean estimated quantity (% of total)
Local Tapwater (drinking, coffee, tea, soup, boiling, baking, etc.)	60
Non-local drinking water (beer, soft drinks, mineral water)	10
Non-drinking water (milk, wine, citrus juice, meat, vegetables, eggs)	30

Table 5. Heavy metals in river water and rain, µg/l.

	Hg	Cd	Pb	V	Cr (VI)	Cu	Zn	Co	Mo
river water, filtered (0.45 µ), global average, (Turekian 1972 (22)	0.002	?	3	0.9	1	7	20	0.1	0.6
rain water, unfiltered, annual average, Wray mires, U.K. 1971 Peirson & al. 1973 (23)	0.2	17	-	4.1	2.9	23	85	0.25	-

Table 6. Lead in waters.

Kind of water		Author	year	ref.
ocean	0.09	Tatsumoto & Patterson	1963	37
coastal, unpoll.	1.7	Abdullah & Royle	1972	38
river, "	3	Wuhrmann	1974	25
rain	34	Lazrus et al.	1970	39
river, polluted	120	Bowen	1966	41
sewage*	200	Mytelka et al.	1973	40

* biologically purified household sewage

Table 7. Intake of lead through diet (48).

Country	μg Pb/day	Authors	year
USA	300	Schroeder and Tipton	1968
Sweden	<450	Dencher et al.	1971
England	200	Tolan & Elton	1972
Italy	550	Zurlo & Griffini	1972
natural	20	Patterson	1965

TABLE 8.

CRITERIA FOR ENVIRONMENTAL POLLUTANTS:

1) CHEMICAL IDENTITY
2) PRODUCTION, USE, EMISSIONS
3) PERSISTANCE IN BIOTA
4) ACCUMULATORY TENDENCIES
5) TOXICITY,
6) INTERACTION WITH OTHER POLLUTANTS

TABLE 9. Uptake, Fate and Action of Heavy Metals in Living Organisms[1]

1. Determination of the chemical forms of metal pollutants in air, waters, food and soil. Study of the mechanisms of absorption and factors which may affect this.
2. Effect of metal ligand binding to proteins on metal transport across biological membranes.
3. Determination of metal contents in critical organs: correlation of toxic hazard with metal content in biological fluids.
4. Development of methods to serve as indicators of subclinical or early toxicological changes, and methods for epidemiological surveys of metal toxicity. Studies on mutagenesis, carcinogenesis, embryotoxicity and teratogenicity of metals.
5. Accumulation of dose-response data and studies of chronic low-level exposure.
6. Studies on the pathogenesis of metal toxicity, correlation of tissue, and cellular and subcellular concentration of the metal with toxicity. Studies on the biochemistry and molecular pathology of metals and the displacement of physiologically functional metals by toxic metals.
7. Studies on toxic interaction between different metals and between metals and organohalogen compounds in living organisms.

[1]According to NATO conference on Eco-Toxicity of heavy metals, Mont Gabriel, Canada, May 1974.

Table 10. Detection limits for Atomic Absorption Spectrophotometer (AAS) (with graphite furnace) and Automatic Emission Spectrometer (AES). Values given in μg/l in aqueous solutions.

Element	Detection limit (in μg/l) AAS[1)]	AES[2)]	Element	Detection limit (in μg/l) AAS[1)]	AES[2)]
As	3	200	Ni	5	5
Cd	0.002	10	Al	0.04	10
Cr	0.4	2	Bi	6	100
Cu	0.1	1	Ca	2	1
Hg	300	50	Fe	0.06	5
Pb	0.04	50	K	0.4	50
V	4	20	Na	0.2	3
Zn	0.001	25	Sb	10	150
Co	2	5	Ti	6	50
Mn	0.02	2			

1) Values reported by Perkin-Elmer Corporation. Graphite furnace, HGA-72, AAS model 403. Volume introduced 50 μl.

2) The AES model Automatic Liquids Analyser 33000 LA by ARL (Applied Research Laboratories).

Table 11. Detection limits of Neutron Activation Analysis (NAA) and X-ray Fluorescence (XF) (excitation by radioisotopes) for instrumental heavy metal determinations of air particulate samples.

Element	Detection limit (in μg) NAA[1)]	XF[2)]	Element	Detection limit (in μg) NAA[1)]	XF[2)]
As	0.04	0.11	Co	0.002	0.09
Cd		0.13	Mn	0.003	0.14
Cr	0.02	0.27	Ni	1.5	0.07
Cu	0.10	0.05	Ca		0.07
Hg	0.01	0.10	Fe	1.5	0.09
Pb		0.13	K	0.08	
Se	0.01	0.05	Sb	0.08	
V	0.001	0.04	Ti	0.20	0.06
Zn	0.2	0.03			

1) From Dams et al. (57). Thermal-neutron flux: $2 \times 10^{12} n/cm^2$ sec.

2) From Rhodes et al. (58). Particulates on 2 cm^2; Si(Li)-detector.

FIGURE 1.

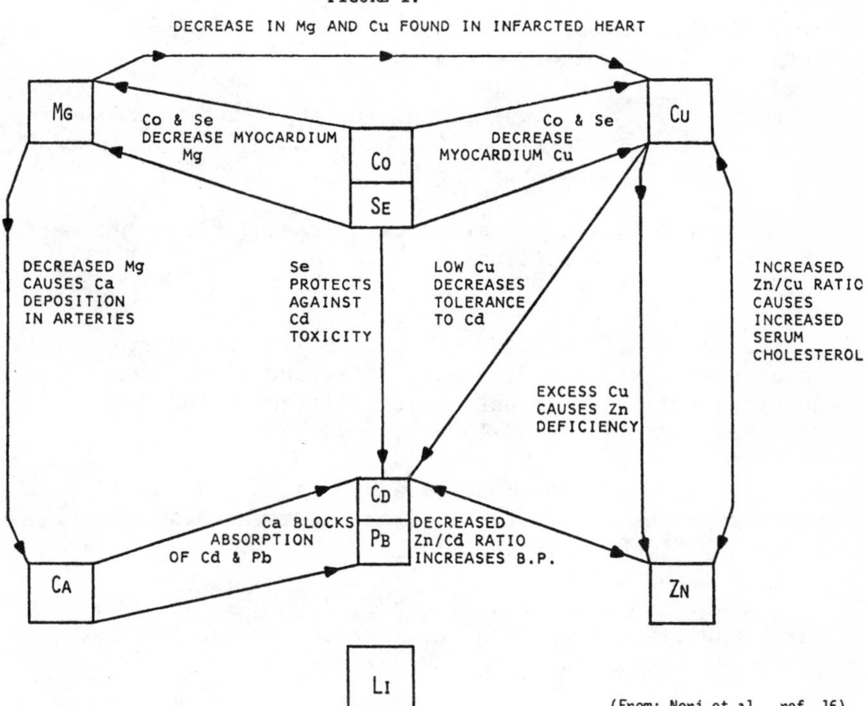

(From: Neri et al., ref. 16)

1. WHO (1973). Trace elements in human nutrition: Report of a WHO expert committee. WHO Techn. Rep. Ser. no 532. Geneva.

2. WHO (1971). International standards for drinking water. 3rd ed., Geneva. p. 32.

3. Rössner, F., Städtehygiene 7, 164 (1967).

4. Benger, H. and Kempf, T., Bundesgesundheitsblatt 15, 17-20 (1972).

5. ICRP (1975). Report of the task group on reference man. Pergamon Press, Oxford. 480 p.

6. Kobayashi, K., Geographical relationship between the chemical nature of river water and death-rate from apoplexy. Ber Ohara Inst Landwirtsch Biol 11, 12-21 (1957).

7. Schroeder, H.A., Degenerative cardiovascular disease in the Orient: II. Hypertension. J Chronic Dis 8, 312-333 (1958).

8. Schroeder, H.A., Relationship between mortality from cardiovascular disease and treated water supplies: Variations in states and 163 largest municipalities in the United States. JAMA 172, 1902-1908 (1960).

9. Morris, J., Crawford, M.D., Heady, J.A., Hardness of local water supplies and mortality from cardiovascular disease. Lancet 1, 860-862 (1961).

10. Biorck, G., Bostrom, H., Widström, A., On the relationship between water hardness and death rates in cardiovascular diseases. Acta Med Scand 239-252 (1965).

11. Biersteker, K., Hardness of drinking water and mortality. T Soc Geneeska 45, 658-660 (1967).

12. Anderson, T., LeRiche, W., MacKay, J., Sudden death and ischemic heart disease. New Engl J Med 280, 805-807 (1969).

13. Masironi, R., Trace elements on cardiovascular diseases. Bull. WHO 40, 305-312 (1969).

14. Neri, L.C., Hevitt, D., and Schreiber, G.B., Am. J. of Epidemiol. 99, 75-88 (1975).

15. Schroeder, H.A. and Kramer, L.A., Cardiovascular mortality, municipal water, and corrosion. Arch. Environ. Health 28, 303-311 (1974).

16. Neri, L.C., Hewitt, D., Schreiber, G.B., Anderson, T.W., Mandel, J.S., and Zdrojewsky, A., Health aspects of hard and soft waters. Paper presented at 94th Annual Conf. of the American Water Works Assoc., Boston, Mass., June 16-21 (1974).

17. Neri, L.C. and Hewitt, D., Review and implications of on-going and projected research outside the European Community, Paper presented at the European Colloquium on "Hardness of drinking water and public health", Luxembourg, 21-23 May (1975).

18. Zoeteman, B.C.J. and Brinkman, F.J.J., Human intake of minerals from drinking water in the European Community, Paper presented at the European Colloquium on "Hardness of drinking water and public health", Luxembourg, 21-23 May (1975).

19. Punsar, S., Erämetsä, O., Karvonen, M.J., Ryhänen, A., Hilska, P. and Vornamo, H., Coronary heart disease and drinking water: A search in two finnish male cohorts for epidemiologic evidence of a water factor. J. Chron. Dis. 28, 259-287 (1975).

20. Karppanen, H., Ischaemic heart-disease and soil magnesium in Finland, Lancet 1390, Dec. 15 (1973).

21. Anonymous (1973). WHO Chronicle 27, 534-538.

22. Liebman, H., Handbuch der Frischwasser- und Abwasserbiologie, Bd. II, Munich, 1149 p (1960).

23. Zemansky, G.M., Removal of trace metals during conventional water treatment. J A W W A 5 66(10), 606-609 (1974).

24. Koppe, P., GWF-Wasser/Abwasser 114, H4, 170-175 (1973).

25. Wuhrman, K., Some problems and perspectives in applied limnology, Mitt. intern. Verein Limnol. 20, 324-402 (1974).

26. Friberg, L., Piscator, M. and Nordberg, G., Cadmium in the environment. CRC Press, Cleveland, Ohio, 1971, 166 p.

27. Murozumi, M., Chow, P.J. and Patterson, C., Geochim. Cosmochim. Acta 33, 1247-1294 (1969).

28. Häsänen, E., Ympäristö ja terveys 9/1975, in print (in Finnish).

29. Yamagata, N. and Shigematsu, I., Cadmium pollution in perspective. Bull. inst. Publ. Health 19, 1 (1970).

30. Jaakkola, T., Takahashi, H. and Miettinen, J.K., Cadmium content in sea water, bottom sediment, fish, lichen, and elk in Finland, Environm. Quality and Safety 2, 230-237 (1973).

31. Miettinen, J.K., The accumulation and excretion of heavy metals in organisms. Symposium on Heavy Metals in the Aquatic Environment, Nashville, Tenn., Dec. 4-7 (1973).

32. Jaakkola, T., Takahashi, H., Soininen, R., Rissanen, K. and Miettinen, J.K. Cadmium content of sea water, bottom sediment and fish and its elimination rate in fish, IAEA Panel-PL-46917, 70-75 (1972).

33. Lucas, H.F., Edgington, D.N. and Colby, P.J., Concentration of trace elements in Great Lakes fishes. J. Fisheries Board of Canada 27, 677-684 (1970).

34. Hartung, R., Heavy metals in the lower Mississippi, in Proceedings of the International Conference on Transport of Persistent Chemicals in Aquatic Ecosystems, Ottawa, Canada, May 1-3, 1974, p. I-93-98 (1974).

35. Hutchinson, T.C., Heavy metal toxicity and synergism to floating aquatics. XIX Congress Intern. Assoc. Limnol., Winnipeg, Canada, 22-29 Aug. 1974, Abstr. p. 91.

36. Patterson, C.C., Lead in the environment, Connecticut Medicine 35, 347-352 (1971).

37. Tatsumoto, M. and Patterson, C.C., Concentrations of common lead in some Atlantic and Mediterranian waters and in snow, Nature 199, 350-352 (1963).

38. Abdullah, M.I. and Royle, L.G., Heavy metal concentration in coastal waters, Nature 235, 158-160 (1972).

39. Lazrus, A.L., Lorange, E. and Lodge, J.P., Jr., Lead and other metal ions in United States precipitation, Environ. Sci. Techn. 4, 55-58 (1970).

40. Mytelka, A.I., Czachor, J.S., Guggino, W.B. and Golub, H., Heavy metals in waste water and treatment plant effluents. Journal WPCF 45, No 9, 1859-1864 (1973).

41. Bowen, J.J.M., Trace elements in biochemistry, Acadic Press, N.Y. (1966).

42. Weichart, G., Pollution of the North Sea, Ambio 2, No 4, 97-106 (1973).

43. Hutchinson, T.C. and Fitchko, J., Heavy metal concentrations and distributions in river mouth sediments around the Great Lakes, Proceedings of the International Conference on transport of persistent chemicals in aquatic ecosystems, Ottawa, Canada, May 1-3, 1974, p. I-69-77.

44. Sukhla, S.S. and Leland, H.V., Heavy metals; a review of lead. Journal WPCF 45, 1319-1331 (1973).

45. Falk, M.R., Miller, M.D. and Kostink, S.J.M., Biological effects of mining wastes in the North West Territories, Techn. Rept. Ser. No CEN/T-73-10, 89 p. (1973).

46. Ravera, O., Lead pollution in air and water, FEG. Informationblatt Nr. 21, 46-54 (1974).

47. Mathis, B.J. and Cummings, T.F. Selected metals in sediments, waters and biota in the Illinois river, Journal WPCF 45, No 7, 1573-1583 (1973).

48. Grandjean, P., Lead in Danes, in "Lead", Suppl. Vol. II of Environmental Quality and Safety, Ed. F. Coulston and F. Korte, G. Thieme Publ. Stuttgart, FRG, 1975, p. 6-75.

49. Laborey, F. and Lavollay, J., C.R. Hebd. Sce. Acad. Sci. Ser. D. Sci. Natur (Paris) 264(24) 2937-40 (1967).

50. Haberer, K. and Norrmann, S., Metallspuren im Wasser. Vom Wasser, Bd. Weinheim, p. 157-182 (1971).

51. Eisler, R. and Gardner, G.R., J. Fish. Biol. 13, 131-143 (1973).

52. Lloyd, R., The toxicity of mixtures of zink and copper sulphates to rainbow trout (Salmo gairderi Richardson). Ann. Appl. Biol. 49, 535-538 (1961).

53. Sprague, J.B., Measurement of pollutant toxicity to fish. Water Research 3, 793-822 (1969); 4, 3-32 (1970); 5, 245-266 (1971).

54. Pietiläinen, Kirsti, Effect of lead and cadmium on the basic production in water. Unpublished. (Master's Thesis in Dept of Biochemistry, Univ of Helsinki, 1974).

55. Pavicic, J., Schultz, E., Korpela, H., Rissanen, K., Laine, P. and Miettinen, J., Unpublished data, 1975.

56. Lindholm, A.-M., Master's Thesis (Dept of Radiochemistry, Univ of Helsinki). Unpublished, 1975.

57. Dams, R., Robbins, J.A., Rahn, K.A. and Winchester, J.W., Anal. Chem. 42, 861 (1970).

58. Rhodes, J.R., Pradzynski, A.H., Hunter, C.B., Payne, J.S. and Lindgren, J.L., Environmental Science and Technology 6, 922 (1972).

ORGANIC CONTAMINANTS IN WATER

Gregor A. Junk

The informal atmosphere at this forum is well suited to what I have to say. The background of the audience is varied and largely unknown to me. Therefore, I have selected some aspects of the organic contamination problem on which I hope to focus your attention. I am prone to make rather blunt and unqualified statements, especially when speaking informally. This may lead to some erroneous conclusions, but I intend to finish my comments in about fifteen minutes. At that time, any misconceptions can be cleared up by open discussion. During this discussion period, we should also have time to pose questions about those phases of organic contamination which have not been considered.

The first aspect of organic chemical contamination is historical and will serve as a backdrop for some later discussions. The measurement of organics in water is a very recent and very evolving technology. In the early 1960's, knowledge of the chemical composition of organic material in water was virtually zero. By 1970, only ten organic chemicals had been positively identified in drinking waters. Since 1970 considerable progress has been made due to concerted efforts to establish the profile of organic chemicals present in drinking water and other waters used as primary sources of drinking water. By profile I mean the list of organic chemicals which have been found in water. Those interested in obtaining this listing may contact me. Relative to my comments here today, it is important only to note that the current list contains about three hundred specific compounds which have been isolated from drinking waters and identified.

However, very little is known about other factors of interest -- the amounts which are present, the distribution, the variation, the source, the control and the possible abatement measures. Even less is known about the health hazards from ingestion of organic contaminated water. I am sure we will hear more about these health effects tomorrow when Dr. Murphy and Dr. Golberg speak to this point.

A few brief comments about these other factors of interest to persons concerned with water quality are in order. Regarding amounts and distribution, the only published body of extensive data is from the EPA National Organics Reconnaissance Survey. The data from eighty different water sites can be used to reliably predict the distribution and the amounts of six halogenated hydrocarbons. Thus, chloroform is 100% distributed in amounts varying from 0.2 to 311 micrograms per liter; bromodichloromethane is 95% distributed in amounts from 0.3 to 116 micrograms per liter; dibromochloromethane

89% and from 0.4 to 100 micrograms per liter. The other three halocarbons, bromoform, 1,2-dichloroethane and carbon tetrachloride, are much less widely distributed and are present in smaller amounts. The point I wish to make by reciting these data is that we have reliable published information on distribution and amounts for only six of over three hundred known organic contaminants. For all other contaminants, with the exception of a few pesticides, the amounts are usually not available or are of questionable validity. Survey data which includes enough water supplies for predicting distribution is also not available. Now, some reasonable estimates about amounts and distributions of other contaminants can be made, but these are not backed by sufficient data. I will elaborate on these estimates later on.

When I mentioned variation as a desirable factor to consider, I was referring to our knowledge of the changes in concentration which occur seasonably or due to changes in waste disposal. For example, the chloroform and some pesticide concentrations will experience a seasonal variation when surface waters are used as raw water sources. The same may be said of contamination caused by waste disposal. These variations, by the way, can be rather dramatic easily covering a range of one hundred or more.

I will not comment on the other three areas of ultimate interest - source, control and abatement. Obviously, knowledge of the source of the contamination is vital to truly effective control and abatement procedures.

Up to now I have stressed the shortcomings in our knowledge of organic chemicals in water. Progress is being made, but one might legitimately ask, "Why has this progress been so slow?" In answer to that question, I will offer a quick comparison between organic and inorganic analyses of water. I hope Dr. Miettinen will forgive me for this cursory comparison. However, the comparison is helpful for developing an appreciation of the difficulties encountered in analyzing water for trace levels of organic material.

By way of comparison water is a difficult matrix for organic analyses, but is nearly ideal for most inorganic analyses. Secondly, there are many thousands of possible compounds in organic analyses, but only about one hundred elements for inorganic tests. (This latter comparison ignores speciation, which is vitally important in both analyses but when speciation is included, the possibility ratio still highly favors inorganic analyses.) A third comparison, background technology, is sadly lacking in organic but extensive in inorganic methodology. Finally, the contaminants in organic analyses are (or were) largely unknown, while the elements in inorganic analyses are obviously all well-known.

This superficial comparison of organic/inorganic analyses helps explain some facts about the measurement of organics in water. The technology for organic analyses is still evolving; the isolation, the separation and the detection problems are horrendous; the results so far, are very incomplete; and finally, the costs are high.

With regard to our knowledge of which organic chemicals are present in drinking water, I mentioned earlier that the number had grown from ten to over three hundred in the last five years. This progress sounds impressive, but the incomplete nature of this listing must be emphasized. Additional components will be added as improved isolation, separation and detection schemes are developed and then employed on real water samples.

Some available data reflect the still incomplete status of the list of identified chemicals. For example, most drinking waters contain about 5 mg per liter of dissolved organic matter. The total amount of individual organic chemicals which are detected by the usual analytical schemes is generally less than 0.5 mg per liter. Thus, 90% of the total amount of organic material present in the water is neither detected nor identified.

While this percentage is suggestive of the unknowns still to be resolved in water research, the situation should not be the cause of undue alarm for several reasons. Firstly, many of the toxic and bioaccumulative components in water have been identified and the amounts are measured using current analytic techniques. Secondly, over 80% by weight of those components which are detected are also identified. A third reason not to be unduly alarmed is that the low 10% identified is based on total weight and therefore it is not exactly related to the ratio of the number of identified to unidentified components. Fourthly, existing evidence suggests that most of the 90% unidentified materials are probably soluble humic and polymeric substances of natural origin. These natural substances might well provide beneficial effects which far outweigh any possible health hazard.

I will now offer some estimates of the amounts of organic contaminants other than the halocarbons for which we have reliable data. These estimates represent a judgment that is made from the limited experimental data which I referred to earlier. Rather than recite numbers for specific compounds and compound classes, I will summarize. The amounts of individual identified contaminants in drinking water are generally less than 10 micrograms per liter (10 ppB). The only exceptions to this estimate are the halocarbons and a very few isolated cases of industrial discharges immediately upstream from a surface source of raw water. Additionally, all identified organic chemicals with boiling points above 150°C (and these represent a good share of the contaminants) are present in drinking water at concentrations below one microgram per liter.

In conclusion, I will make some comments which I feel should be of interest to people concerned with water quality. Most of these comments are defensible based on existing scientific evidence. Others are merely my opinions as a water researcher who is concerned, but not alarmed, about the deterioration of the quality of water due to the presence of trace amounts of organic contaminants.

My first comments are directed toward methodology or the procedures used to measure organics in water. No simple measurement is available for routinely checking the organic contamination of water caused by compounds of suspected or known toxicity. The more simple tests, such as dissolved organic carbon, total organic carbon, biological oxygen demand, chemical oxygen demand and the recent spectroscopic methods <u>all</u> fail to provide a valid indicator of drinking water quality. Eventually, a gross indicator test, analogous to the fecal coliform test for bacteria, may emerge, but no such simple test for toxic organic materials is now available.

The somewhat encouraging correlation between the relatively simple test - total organic carbon versus the amount of halocarbons in water - must be viewed, at best, with guarded optimism. Future results may establish the validity of this correlation. Then we would have a valid test for six contaminants. However, a serious problem would still remain unsolved. What about the other three hundred chemicals known to be present in various water supplies? It is extremely doubtful that the concentrations of these compounds will parallel the total organic content in the water under test.

At present, the only accurate indicator of water quality involves the isolation, the concentration, the separation and the identification and measurements of individual organic components. Such procedures require trained personnel and are very costly.

It is unrealistic to expect water purveyor or even state water quality laboratories to do regular or even one-shot analyses of drinking water for all the possible organic chemicals of interest. Even with highly trained and experienced analytical chemists using expensive instrumentation, no single methodology is capable of analyzing for all the contaminants known to be present. Aside from cost factors, the potential for this development in the foreseeable future is almost nil.

A second quick comment about the current list of organic chemicals present in drinking water. Polynuclear aromatic hydrocarbons are rarely reported, especially in U.S.A. drinking waters, because of the failure of the methodology employed. These aromatic hydrocarbons are most certainly present when surface waters are used as raw water sources. This is but one of many examples which could be cited to substantiate the incomplete nature of current information about the presence of organic components in drinking water.

Now, a brief comment about recently publicized technology for removing organic chemicals from water. Any procedures which remove _all_ the organic carbon from water may be ill-advised. Selective removal of those chemicals which have known or suspected toxicity is certainly more desirable.

My final comment is concerned with health effects, possible regulatory standards and control measures. These must be considered from a broad spectrum of inputs. I will cite but two of many possible examples which illustrate what I will call _equivalent ingestions_. Some may prefer the term, _total body burden_ as was used in some discussions yesterday. In the State of Iowa, the regulatory limit for dieldrin in milk is 300 micrograms per liter. Based on an extensive survey of the dieldrin level in drinking water, we find that 15 nanograms per liter is a probable upper limit. Thus, from one quart of milk, a person can ingest an amount of dieldrin equivalent to that present in a lifetime (70 years) supply of drinking water at the rate of two liters per day. Another example in case you are not a country boy who likes his quart of milk a day - suppose you eat a one-pound fish which has matured in water containing only 10 nanograms of dieldrin per liter. Because of biomagnification in the fish, you would be ingesting an amount equivalent to that present in a 20-year ingestion of dieldrin contaminated water. Many other examples could be cited where alternative routes of ingestion result in a far greater human body burden than that occurring from ingestion of drinking water. It seems foolish to spend time and money removing or stringently regulating certain known contaminants in drinking water when other environment and life factors cause a far greater exposure to the same potentially toxic chemicals.

This last comment is an attempt to place the ingestion of organic chemicals from drinking water in a reasonable perspective. I am _not_ advocating abandonment of regulations, scientific study, and concern about the presence of organics in drinking water. Rather, I am suggesting that firm scientific evidence is still missing; health effects are still largely unknown; and additional regulatory actions at this time would be premature.

DISCUSSION

DR. SMEETS: I think we might have a very interesting discussion. Dr. Junk, your paper gives me something to think over. That means you are suggesting you can identify quite a number of organic contaminants in water. But the point is, how far can everybody in practice do this kind of control? That is the point. I remember the story of the analytical chemist in one of the institutes in Holland, the Institute of Public Health, who discovered a strange peak on a gas chromatograph, and he discovered that it was a chemical that came from the other side of the frontier.

And, he didn't know what it was, and it was considered at the moment to close the drinking water supply to Amsterdam. No one knew where this came from, but the man in the laboratory was very alert, he saw a foreign peak in the gas chromatograph, he discovered it was a dangerous pesticide. So, I mean there is a difference between practice and research, if you let me put it this way. And this is a very important point. Maybe we can come back later to this, but anyway, in the River Rhine, quite a lot of fish were killed, and we do not know what should have happened if the analytical men in the laboratory did not have discovered their foreign peak. But we might come back to this later.

Since we have had three very interesting papers this morning, of which the subjects are pretty similar, I suggest that we follow another procedure other than that of yesterday.

You remember that yesterday we had discussion referring to each paper separately. I am suggesting today that we should have a panel in the forum. I would suggest to everyone here who wants to raise questions, that he do it in one of two ways. He can ask a question directly to one of the panel members, referring to the paper presented today, or he can ask more general questions, referring to a topic which was not dealt with this morning.

Before starting the discussions, I would like to refer to a point already mentioned earlier this morning. That is the quality control program and the comparability of measuring data. To provoke the discussions, I would like to cite, in text, from a paper presented by Mr. Thatcher of the United National Environmental Program. This could stimulate a little bit of discussion, since I do not completely agree with him.

It is of interest to read it to you, and he says, "About two years ago, the Monaco Laboratory, which is responsible to the International Atomic Energy Agency, in the calibration of measurements or radio nuclides, in a marine environment, performed an experiment with a common sample of sea water. Fifty of the world's leading laboratories sent in their results .

And the published results indicates that no one in the world knows how to measure radio nuclides in the marine environment in a way that is compatible with anyone else's measurement. If that is the case, with radionuclides, in some of the other polluted categories, it is obviously even more serious." I was just quoting this particular point since I think the comparability of measurement data is one of the very important points we might discuss this morning on the panel.

The floor is now open to raise questions, or to make general statements, or to ask more particularly with reference to the papers which have been presented.

DR. TRUHAUT: I would like to make a general comment. We listened this morning to excellent reports dealing with problems of analytical chemistry. That is a problem of control. And this shows how important it is to have a close liaison between the toxicologists and hygienists on the one hand, and the analytical chemists on the other.

Because if biology recommends some value not to be exceeded, it might be a disaster if there is no adequate analytical method available. It is very important to have this close liaison. And because of that, I take the liberty now, of making in the beginning a general comment relating to toxicological problems. That is, the establishment of water quality standards.

In my view, this publication of values expressing concentrations in micrograms per liters is something that could be dangerous, if we forget that overall we have to look at water composition in what I will call an integrated concept. That is, in context. This is what we cannot ignore. Dr. Miettinen stressed that yesterday and again this morning. He stressed how important it is to compare, to take into account what is coming from other sources than water. This is my first comment.

My second comment is, that instead of establishing concentrations, or even for establishing concentrations not to be exceeded, we have to rely on the concept of acceptable daily intake. But in the case of water, and especially in the case of inorganic pollutants, I think that the concept of acceptable daily intake must be fruitfully replaced by another one, that is, a concept of a weekly intake. Because really, the consumer's ingestion has to be looked at toxicologically over a long term; not in the immediate context, you see.

For example, ingestion of mercury or cadmium, or lead, has to be looked at for a long period of time. I hope I am clear, in my broken English, but it seems to me that this is very important to take into account in looking at the analysis and safety of drinking water for control purposes.

DR. SMEETS: I think this is the way WHO and CEC is already dealing with the problem. They are changing their minds from the ADI (Acceptable Daily Intake) to the weekly intake.

DR. TRUHAUT: I have another comment to make to stimulate the discussion, after which, I will keep silent. It was mentioned by one or two speakers this morning that for some inorganic elements we have to note that some low doses may have a beneficial effect, while higher doses might cause a harmful effect.

I will refer to the case of selenium. Without selenium, the calves, for example, have poor muscle function. In poultry, there is also a disease caused by deficiency in selenium. And in man, you know, selenium is very useful for liver function. I will not elaborate on this, but what I will stress is that selenium, at very low doses, is essential. At higher doses, it may cause liver cancer. And here we have a very spectacular example of what I personally call a secondary carcinogen.

We have a spectacular example of a secondary carcinogen for which it would be completely foolish to say that there is no threshhold. There is a threshhold, because there is a need for selenium. This is very important; the dose-effect relationship has two parts; first the beneficial effect, and then second, the carcinogenic effect. And you see how it is sometimes asientific. Ascientific means that there is no - not at all - no scientific value to say that there is no threshhold. I wanted to mention this before the panel today, this completely foolish idea, to stimulate discussion.

DR. SMEETS: With respect to your integrated monitoring, Dr. Truhaut, I think we can all fully agree with the idea. And, more or less, I have been referring to this this morning, talking about lead. And maybe Dr. Miettinen wants to answer the problem raised by you. I should like to give the floor to Dr. Miettinen.

DR. MIETTINEN: Well, we do not want to talk only on lead. To complicate a little more this question of selenium, I could mention that selenium protects against mercury and cadmium toxicity. With a higher selenium level, the toxicity is not reached at similar levels for other heavy metals in water. On the other hand, selenium also increases the intake of these heavy metals so that the fish and other sea animals, if there is both selenium and mercury or cadmium in the water, have higher levels of both. But at least toxicity-wise, does not become visible at the same levels of both.

How is the cancerogenic effect produced? I don't really know. Now, about lead. I know that Western studies show that there is not so much lead in the atmosphere, but I also know other studies where lead from motor cars is present in urbanized people from inhalation. They also get lead from other sources, as has been shown by isotope radioactive studies. For instance, in Finland, the soil contains a ratio of about 14.5 times the radioactive change, radiogenic lead 208, rather than the rare isotope 204.

But most of our gasoline comes from the United States, and 85 percent of lead in the U.S. gasoline comes from Missouri. And it has the higher ratio, about 20. So that all lead in gasoline is labelled, which is used in Finland. We can differentiate very well the lead in man or the influence of the gasoline on lead in man. And such studies show that about one-third of the lead originates from gasoline; two-thirds originate from other sources.

Our studies in Helsinki show that about half of the air pollution in Helsinki originates from gasoline, the other half originates from mismanaged smelters and incinerators. And all of the air concentrations are not directly perhaps the main source. All we eat is also influenced by the lead in air. It is vegetables grown in polluted atmospheres, and the canned vegetables. Even more comes from the lead soldering and drinking water, in some areas, from lead tubing, and so on.

There are, of course, big differences, but mankind is more or less under a lead burden from technology. And it is known that lead in the blood of urban populations has increased. It is about two-fold the natural level. And the margin for biochemical changes in blood already has been lost in rather large populations where some occupational groups have about 25 or 30 micrograms of lead per 100 grams of blood, which is just the level where the alpha level is disturbed. We don't know if this change has any harm to man, but usually these changes are not beneficial. And the margin of safety of lead in blood to the lowest clinical level is not very wide any more. It begins at about 60 micrograms per 100 grams of lead, or so.

DISCUSSION NOTE: There followed much talk about the safety of lead in the air, questioning the statements made by Dr. Miettinen. Dr. Smeets ruled from the Chair to postpone this discussion, stating he would prefer at this time to stay with the topics of the morning.

DR. KORTE: Dr. Egan, where do you see the future development, the main development, in standardization of packaging methods and development of patented group methods, or in single chemical analysis?

The second question I would like to ask you refers to

your long time experience in analytical chemistry. What do you think is more successful? To standardize analytical people or to standardize the methods?

DR. EGAN: In some ways, the answers to both questions from Professor Korte are related. But, he asks first of all, do I see the future development more in the sense of a packaged index, which relates either to a number of different individual indices, or to what I might call an artificial index, or on the other hand, whether I see it in relation to single chemical entities. I think that is the question.

I can only answer that by referring back to what I said fairly early on in my presentation, and that was this. That the analytical chemist is not on his own in this field. He is really working with the toxicologist, with the marine biologist, with the plant pathologist, and all the other environmental biosciences, which relate. And the chemist cannot individually give guidance here. If lead is important by itself, if mercury is important by itself, then those analyses will continue to be required and to be important.

On the other hand, when we are talking about total organic loadings, it may be quite proper again that we discriminate individual organic compounds. Dieldrin was mentioned this morning. I would not necessarily cite that, but there might be a number which should be looked at individually. On the other hand, there might also be great value in compiling an arbitrary index which has some biological significance, some toxicological significance. And that may vary from one medium to another. It may be one thing in water, it may be a different thing in air, it may be a different thing again in food. But, I think, there is room for what I might call arbitrary indices, provided they have some correlation with biological phenomena.

Then you ask, Professor Korte, whether I think it more important to standardize the analytical method, or to standardize the analyst. I think you've got to have a competent analyst to begin with, whether you've got a standard method or whether you are going to leave it to his discretion what method to use to measure selenium, for instance.

But, there are some concepts that are necessary to standardize on paper, because unless you do that, you won't measure the thing that you are setting out to measure. The classical example is not in water, but it is in food and animal feeding stuff analysis, where one has an interest in the fiber content of the feed. Fiber is defined by the actual process you use analytically to measure it by. And it is necessary, therefore, to write it down on a piece of paper, for everyone to do the same analysis.

Because that is what fiber is. There is no other definition of fiber. On the other hand, if we are talking about lead, which is an element, a chemical entity like DDT, then there is some scope for professional competence to decide which particular approach to use. And this almost brings me back to the question raised previously. There are many well-developed schemes for establishing standard analytical methods, and most of them - not all of them - are based on collaborative studies in which, as here, a single sample is circulated to very many analysts.

What is necessary, and what has, in fact, developed in many cases, is a systematic way in which one evaluates the results that are obtained in that way, as to whether they are sufficient data to establish a standard method or not. There is a systematic way of deciding how many analysts to involve, how many different samples that each test, or how many replicates they should do, and even how to evaluate the results, the validity of them.

It is that framework, that system, that philosophy of collaborative study, which needs to be fully recognized, and no wonder - in an initial circulation of samples to fifty laboratories, there is a lack of agreement. But there is a need, as I say, to recognize some guidelines for establishing collaborative studies. And there may also be need to establish reference methods, too.

But, perhaps I oughtn't to leave the subject without referring to standard reference materials which are, in a way, the backbone of collaborative studies. Now, I don't know whether I have fully answered Professor Korte, but I have tried my best.

DR. JUNK: My comments here will be somewhat related to what Dr. Egan just mentioned, and that is, the possibilities for group analysis in organics. But, before that can evolve, we will need help to develop some type of screen which separates the toxic from the non-toxic materials. If I could just recite again some figures. If the total organic load is of a level which you can measure, which is normally, for most of our instruments, a break-off at about one part per million, the levels that are usually present in water are somewhere of the order of one to ten parts per million.

Now, if we decide that we are concerned about specific organic chemicals being present in the part per billion level, then it is naturally impossible to try and take the difference and inaccuracies between two large numbers and arrive at a very low concentration, and say, we have now some type of group analysis, in this case the total organic carbon, which is indicative of the changes in toxic level in that water.

That may evolve, but right now we do not have this screening program for doing this.

DR. KORTE: If you discuss organic chemicals in water, or some other medium, we always have difficulty not to know what we are looking for. Did you try to develop any procedure to predict what to look for, or something like that, during your studies?

DR. JUNK: Yes, initially in trying to develop procedures, we were just trying to find something. As this evolved, and this is related to a comment which Dr. Korte made in his presentation yesterday, one gets a pretty good idea of what to look for, by just looking at the production levels for organic chemicals. So, you get a pretty good clue as to what you might find in the water environment by looking at production levels.

Now, another thing which I think was hinted at in your talk yesterday, is you will also look at the persistence of that particular organic chemical. If it is very persistent, your chances of finding it in the water, or in a living species, which survives in that water, is fairly high. Because usually it biomagnifies in the food chain.

DR. KORTE: Where do you get the production figures for organic chemicals?

DR. JUNK: Amazingly, if one inspects the tariff figures in the USA, the recrods on tariffs, you can get a pretty good estimate there. Now, you cannot always do this with every chemical, but by and large, you get a pretty good estimation of production levels there, and I don't know if this is amazing or not, but we find that we get very good cooperation from industries. Where we cannot locate the figures, we often find that if it is a competitive chemical, the industries themselves know who is producing what amounts, and you get very good cooperation from the industries.

DR. BLAIR: I think this is important, because you are talking about the production figures. Yet, in our discussions, I raised the question about bromoform, which was among the six most widely distributed. And bromoform is really not a compound of commerce, you know. It must be a piddling thing, in size, and your answer was, I think, to me, most interesting. And I think the group should hear about that, too.

Of course, chloroform is a product of commerce, and widely manufactured. But bromoform, no.

DR. JUNK: I will address myself, first of all, to the chloroform level, and what we do know now, which has not been verified by independent laboratories. I think some of the best work on the chloroform levels in drinking water has been done in Amsterdam. It appears as if the chloroform is formed by a reaction between the chlorine used to disinfect the water and the soluable humic material, which is present in that water.

For a while, people were speculating that, well, how can we possibly get bromoform in that water. And it was suggest that it is probably an impurity in the chlorine. The bromine level in chlorine is extremely low, and you cannot account for it being a reaction of bromine on the humic material. But, if you look at the kinetics of that kind of a reaction with the humic material, you find out that if there is bromide present in that water, which there usually is, the kinetics favor the formation of the chloroform, and then, an exchange of the bromine for the chlorine, so you can very logically explain the high bromoform level. And I well agree, if you inspect the production levels for bromoform, you would never predict that the bromoform concentration in drinking water is going to be, or in any water, is going to be very high.

I would hasten, then, to add that the bromoform concentration which is in the treated water, is not in the raw waters. If you look at the raw waters, indeed, it is a rarity to even find chloroform, where there is a significant production.

DR. COULSTON: I would just like to remind the group that chloroform, which has a lot of prominence, now, is used quite widely in cough syrups, expectorants, and toothpaste, as a flavor enhancer. The amount allowed can go as high as three percent, as I remember. I don't understand what all the fuss is now about finding a little chloroform in water, when we've used chloroform in all kinds of products for at least 75 years. I am not arguing for or against chloroform. I just want to make a point about a chemical that has been around for a long time, without any evidence of its harm to man.

DR. MRAK: I thank what Professor Coulston said about chloroform is a good example, and it is very well taken. I would have liked to make a similar remark, and give a question to Dr. Miettinen relating to the hardness of drinking water. If we look at people and all these measures are taken on behalf of improving the health of the people, what is the most important disease these days? I think most people would agree that it is cardiovascular disease.

Now, Dr. Miettinen said that a correlation was found between the distribution of elements in infarcted hearts, in hearts which have suffered a coronary occlusion, and the hardness of the water. My first question is, how do we know, or why do we suggest that this correlation is a causal relationship? Question, could it be that in the hearts that have been diseased by impairment of the muscle, by an insufficient circulation within the muscle, that the metabolism of the cells of the heart are changed, which would cause the differences in distribution of the elements? Has this been looked at, have animal experiments been done? It will be very easy to do that experimentally, by making a mechanical coronary occlusion and see if that was or was not followed by a different distribution of elements. It is just a question.

And another question, is on the difference in hard and soft water areas and human and cow's milk. How is it in those areas that the elements in human and cow's milk differ between hard and soft water areas? I feel this is a very important and pertinent question, because we know that cardiovascular disease starts almost before we are born.

DR. SMEETS: I have already referred yesterday to the recent colloquium. We had a lecture on this particular subject and invited three of the best Western European nutritionists. We asked them if they could inform us about the following point: soft drinking water has a lack of certain metals, like calcium, magnesium, and so on. And, we wondered whether such a lack of special metals, like calcium, magnesium and so on, would not be compensated by their presence in food. We were considering the compensation process.

I must say, we had a great fight between the nutritionists. They did not at all agree! Some of them said that it was a compensation of the food, for physiological and metabolism reasons. I think it is a very difficult point. There are quite a number of questions, and I might come back to this particular point tomorrow, but now I want to give the floor to Dr. Miettinen, referring to his work.

DR. MIETTINEN: I don't really know what to answer, except that I did not say there was a correlation between the metals in the drinking water and heart disease. On the other hand, I stressed rather that the causal relationship is not proven, although the correlation is real. There is such a correlation; a question of debate among the nutritional scientists, whether these elements are really involved, or whether the correlation happens to be for some other reason.

It seems to be very complicated, but I felt obliged to mention this briefly in my written paper, and also in the oral presentation, of the large number of studies which have been made on this problem. And, maybe there is some relationship, because in Eastern and Western Finland, there seems to be really very marked correlation, which makes one wonder if there isn't something to it, although the data as such doesn't prove anything about the cause. Maybe that is enough.

DR. SMEETS. I wonder whether Dr. Egan would refer to the studies made by Crawford or by Shaeffer in England, because in England, they have done quite a number of very interesting studies to which I referred yesterday, considering the north-west-southeast line. I wonder, Dr. Egan, if you want to refer to these studies.

DR. EGAN: I think possibly we are getting outside the main area of our discussion. In Britain, the situation, to me, is a little bit complicated by the fact that we already add calcium carbonates to all flour, except for minor usage, and that is whole meal flour. So, I find it rather difficult to make a constructive comment in any case on the situation in Britain. I think it is fair to say that there is no single agreed view of the situation, but I think in Britain, we may have even a more complicated position.

DR. CHICHESTER: I would like to make two comments. One, to reinforce the earlier speaker that we, as nutritionists, still are not aware of all of the requirements, if you will, particularly of the trace elements, for the well-being of man. I would like to reinforce the comments pertinent to the interaction of various elements. Besides selenium and mercury, you have, of course, copper, molybdenum, vanadium, lead, calcium, all of which interact with one another in that they change the toxicity of one element with the ingestion of another.

And, since we don't know the requirements with any exactitude for many of the rather micro-trace elements, we still must consider, from the nutritional standpoint, what the results would be if one absolutely removed all of these elements from drinking water supply. And I think one must keep this in mind when one discusses the trace elements.

But additionally, and since I am also a food technologist, one of the more interesting aspects which is left out of the consideration of water quality is the acceptance of the water to the population. I think you will recall some work that was done a few years ago on the acceptability of water, drinking water, on the basis of its mineral content. And, I suggest that perhaps in the correlation one sees between the mineralization and various other disease states, that the acceptability of the water supply may influence greatly the consumption of the water.

I think, also, that this is an aspect of water quality

which should be kept in the minds of the people, and discussed, rather than exclusively on the toxicology of the elements.

DR. GREIM: When Dr. Egan discussed analytical standards of water quality, he also attacked the problems of waste water. But how about the waste water having passed the sewage plants? Are there any comparative standardizations in the different countries, or can one at least compare the different procedures used in different plants in different countries?

DR. EGAN: I don't know that I can answer the question satisfactorily. I think there is still room for comparative results. It doesn't necessarily follow that there has to be a great standardization all over the world. It is important, from the health point of view, which is properly understood regionally and locally, and there may be, of course, a question of importing and exporting water from one country to another, across a national barrier.

But given that one has a reasonable understanding of the health implications, unless there is some import-export situation, it may not be necessary to have a 100 percent identical control throughout the world. There are regional and local matters which may call for different approaches in some, if not all, of the aspects.

DR. KÜHN: Dr. Egan, I presented a paper last week in Las Vegas, about quality control, and I have a question. I was very surprised when this question was raised from the floor: one of the people asked me that when you approve of all your measurement techniques, how can you compare, then, the new results with the older results? I must say, my answer was to this point; if you want to stop improvement of measurement techniques, I think those questions are a bit strange.

DR. EGAN: Yes, there may be need to continue using some of the old indices in order to have some kind of understanding between the interrelation between the previous standards or levels of quality and the present ones, but I think that this probably will continue in some areas for some time. But that should not be to the complete detriment of creating new standards or indices, if these are more valuable. If they are better related to health or water quality, or whatever.

MR. GOULD: Perhaps we are coming a little bit to a conclusion here, that drinking water per se is perhaps not a major carrier of materials into the nutrition of man. And perhaps what we are saying, in effect, when we are talking about contaminants, is gross contamination of water. There are a lot of traces, the significance of which we can debate indefinitely. However, one of the factors we are concerned about today is to prevent the gross contamination of water by organic chemicals and metals.

I think, to put a new perspective in here, as one who some years ago actually was involved in the operation of water treatment plants, as far as the public is concerned, they have a somewhat different perspective on drinking water. They are concerned about, as was mentioned a moment ago, the aesthetic considerations of odor, taste and appearance. And, I think we also have to be concerned; those that drink this water in Washington could not fail to notice a chlorine taste in the water.

We have a problem here which is that the standard of drinking water in many parts of the world is below optimum on an aesthetic basis, quite apart from the dissolved contaminants. And the dissolved contaminants, in many instances, are only of significance when there is gross contamination from a polluted source.

Now I would like to comment on two or three statements made this morning. One is the statement made with respect to mercury, of the rapid methylation that takes place. Now this may be a laboratory observation. However, we have done some work in the vicinity of a plant which was discharging some five kilo of mercury into water over a period of some twenty years. The area is an ocean discharge, and the highest concentrations of mercury that were found in the fish, and these measurements were conducted by the Food and Drug Administration, showed a concentration of 0.6 milligrams per kilogram.

There were large concentrations of mercury in the sediments, and examinations of these sediments, which are very biologically active, showed only traces of methyl mercury. Shellfish examination showed only traces of biological magnification. So it is agreed that this process can take place; however, there is a very substantial dispute as to the rate at which this process takes place in nature as opposed to laboratory conditions.

There has been talk of nitrosamines in water. Bear in mind that many natural waters, and some of them from chalk wells, are high in nitrite and ammonia. And, it is not very far from nitrite to arrive at nitrosamines. There is never a natural source of nitrosamines, which I think might well be worthy of examination.

I have one concern in the discussions, and that is, we have a tendency to jump from talking about drinking water one minute, to surface waters the next minute, and sea water in another minute. Really, we have three different scenarios, and some of the remarks we are making would be interpreted differently in each of these three areas. Just to take an example on analytical methods, let's take one simple test, suspended solids. The conventional analyses for suspended solids, even under so-called standard methods that we use here in the United States, show great variability, even when conducted with the same analyst, due to the very heterogeneous nature of the solids themselves.

We have a problem that we are approaching here, where regulations are being set which are very stringent, which are being applied across the country, that if one cannot place faith in these analytical methods, then what legal basis can there be for enforcement of some of these numbers? So, we have a very significant problem of doing the best job we can to make these analytical methods as effective as possible.

And now, for example, on TOC. Now the significance of TOC would be very different, for example, in a treated drinking water as compared with surface water, where you may have large quantities of cellulose fibers. You could have, for example, in a biological treatment plant where you are discharging biological solids that their behavior in passing through the conventional filtration medium is quite different from that of mineral solids, which we are normally accustomed to handling in a suspended solids determination.

With the rapid growth of secondary treatment plants, we are seeing a significant problem here. Firstly, how do you handle this problem of determining suspended solids; and secondly, what is the significance of these suspended solids, these biologically suspended solids, when you have measured them? What happens to them when they go out into the surface water, into the food chain? These are some matters which occurred to me from the discussions this morning.

DR. MIETTINEN: Yes, there are indeed great differences in the rate of methylation. We have studied it in laboratory, and found out that in aerobic conditions with different samples, about one-third of the mercury content is methylated in about two weeks. This is just one example. But, in anaerobic conditions, there is usually H_2S present, and then, mercury forms a very difficult soluable sulfide, and is not methylated; so there can be any rate in nature. It just happens that in many lakes, the methylation seems to be relatively fast in the aerobic state.

DR. EGAN: Dr. Miettinen has said some of the things I wanted to say about mercury. I would like to make the additional point that because of the different rates of methylation and different circumstances, the equilibrium proportion of methyl to inorganic mercury may differ from one substrate to another.

That is to say, for example, you might find that 30 percent is converted in mud, 85 percent in fish, in equilibrium in the same situation. I don't know whether Professor Miettinen would agree with that, but it doesn't follow that you have 90 percent conversion in the mud, that you must have 90 percent conversion at equilibrium in the fish in that locality.

Regarding nitrosamines, the kinetics of nitrosamine formation between nitrites and secondary amines has been studied quite extensively, and the conditions for the formation of nitrosamines really relates to a rather narrow pH band. I think it is not conducive, in general circumstances, for nitrites automatically to form nitrosamines. There is room for discussion here, and we heard some preliminary observations yesterday, which to me, require much fuller confirmation. But, I don't think one should automatically assume that because nitrites and secondary amines are present together that nitrosamines are inevitably going to be formed.

On the third point, which Mr. Gould made, he referred to national, regional, local and other criteria not necessarily being the same. He asks what faith one can have in analytical methods. I would simply like to repeat the importance of interlaboratory studies and the conscious exercise of laboratory control, quality control, based, for example, on standard samples or standard reference materials.

DR. JUNK: I don't know whether I should comment on the question or the comment, so I will do both. Referring to the comment, and I'll use dieldrin as an example of equivalent intakes. I was glad to hear yesterday that the EPA was cooperating with the FDA in attempts to set interim and eventually final standards. And chloroform was mentioned, as in cough syrup. One can arrive at many examples of equivalent intakes, polynuclear aromatics by inhalation and gasoline aromatic hydrocarbons.

I am of the opinion that if we look at equivalent intakes and what is known now about the toxic materials which are present in water, we would have other forms of human ingestion of the very same materials, which would far outweigh what we ingest in water. I don't mean to leave you with the impression that we should be unconcerned about what we ingest in water, but I do think that the other intakes of the same chemicals are far greater by other routes.

And regarding the TOC question which was raised, I agree it is rather arbitrary, but within one laboratory in the USA, arbitrarily we define it as dissolved organic carbon. Most everyone could agree that dissolved organic carbon is a better parameter than total organic carbon if it passes as .45 micron filter.

Now, some suspended material passes a .45 micron filter, but that is sort of the arbitrary definition of what is dissolved in that water. If one then makes the measurement of dissolved organic carbon, obviously in Germany on the Rhine River, they have been able to make some very accurate measurements, because they are able to take the difference between this and make correlations between the biodegradable material and the material which is naturally present in that water. So I don't know if we are that bad off in the area of standard analytical techniques.

DR. COULSTON: The International Academy of Environmental Safety has published a special volume on lead, in the periodical "Environmental Quality and Safety." In that volume are many papers on lead which refer to Dr. Miettinen's statements. There are two papers by Dr. Travis Griffin and his colleagues which present the work done in my Institute on lead, given by inhalation to rats, monkeys and man. I think the answers to some of the questions are discussed in this volume.

MR. DONALDSON: I think I would like to put some things back into a little bit of perspective here. This is an international forum on water quality, but yet, most of the discussion that has taken place in the last two days has really only covered about 30 percent of the world's population. Most of this has centered about Europe, and when you look at graphs of the world's population, you will find that what we are talking about is really a relatively small portion of the world.

There was one other statement made, that I don't think I can really let go pass, because of what I have just said, and it was one that was just said that heart disease is the disease that affects most people in the world. I think those of us who have spent much of our life in Latin America and elsewhere will take great issue with that; parasitic and bacteriologic diseases are the leading causes of killing people.

I can give some figures. In the countries of Latin America, which happen to be my area of work, the mortality rates in the age group of under one year range in the order of 2,500 per 100,000, compared to 27 for the USA. The rates in the 1 to 4 year group range are in the order of 600 to about 300 per 100,000. And just for comparison, the USA is 1.8 per 100,000. In other words, we are talking rates almost 700 times what we are talking about here in the USA.

So that when we spend a lot of time talking about whether a man is going to die of mercury poisoning, it is for much of the world, a very academic problem. And, I would like, and I would hope that there will be some additional discussion on these other problems, which are very real problems. The problems that we are discussing are very real to all of us, because that happens to be the world in which we live. But I think we have to put the discussion in that context.

There is one other thing Dr. Egan brought up, and it was, again metnioned on several other occasions, that I would like to touch on a little more: it is the question of multi-level standards. You point out that WHO has worldwide standards. They have, since that was published, come out with a European standard. There was currently a meeting in Geneva in which they are now talking of five levels of standards. I would be interested in a little more discussion of that point.

DR. SMEETS: I am very impressed by your statement, and you are, in certain measure, completely right. But the question is always at an international symposium, how far is this internationally? For instance, I know Greeks, I know Italians in this meeting, and do we speak for them? We do not speak for them! As I learned yesterday from the President of the International Academy, a meeting is foreseen next year in South America, so the typical problems for those regions might then be discussed more fully.

DR. COULSTON: Mr. Donaldson, I think you are absolutely right, and of course, that is why you are here, to present information involving Latin America. And we have another speaker who will represent the Central European countries. We have a representative from Japan, who will talk about the Far East. We don't have anyone from India, and we don't have anyone from Asia Minor, but we have covered as much as we can of the world.

I am just saying this in defense of the organizers. But, sir, that is why you are here, to make the statement you just made. And I thank you, for all of us.

DR. EGAN: I don't think I can add very much more to what I have said already in my presentation. I said, quite simply, that analytical methods of pollutants in the aquatic environment are of interest on a local as well as a regional national and global basis. And by that, I mean to convey, in complete agreement with Mr. Donaldson, that one does not talk about standards in relation to one region only. And I did myself draw attention to the two different standards of World Health Organization.

I think we touched upon this yesterday, even in the more commercial sense of bottled water, and the fact that some people, perhaps in New Zealand, or Sicily or Mexico, would require a trace of sulfur, or even selenium, in their water to make it acceptable.

DR. RICE: I am the technical advisor for the International Ozone Institute and I would like to back up to the bromine in the drinking water and some recent data that probably explains some of the brominated compounds in the drinking water. It has been known to the ozonatics of the world for some time that if you ozonize sea water, which has in it some 35 ppm of bromide, that you can very easily oxidize the bromide to bromine. And, if you ozonize sea water, you will see bromine coming off of it.

The theory was then discussed at our second international symposium on this problem in Montreal in May, that some of the ground water might have bromide in it, and when it is then treated with the more powerful oxidizing agents, namely, elemental chlorine, you would also form free bromine. That was just a theory until about a month ago, when I saw some data from the EPA Kansas City region, in which they took whatever river water in that area, and did the analyses for the six hydrocarbons, then added some KBR to the point of 1 ppm of bromide.

They had that go through the normal drinking water treatment process which ended up using elemental chlorine, and they came out with at least one order of magnitude more of three of the six brominated compounds.

So that doesn't necessarily prove the source of the bromoform, but it does mean we are probably manufacturing some of the compounds in our drinking water process. And, I would like to make a second point, and that has to do with the ozone layer above the earth, which you might not think has anything to do with drinking water, and it may not.

As you know, the question that has been raised on the upsetting of the total amount of ozone using the argument of a free radical, namely, the chlorine radical. Recently, Professor MacElroy of Harvard has also volunteered that if there were bromine free radicals up there, that these would last longer and do more damage to the ozone layer. A final point is that Dr. Roland at the University of California at Davis has recently put out the theory that if a halogenated hydrocarbon has hydrogen in it, it will not rise above the trophosphere without being decomposed. But if it has no hydrogen on a carbon, then that halogenated carbon compound can rise through the trophosphere into the stratosphere. And, if you look at some of the six brominated, chlorinated compounds in our drinking water, some of these have no hydrogen.

And so, I just throw that out as a possible interrelationship.

DR. MIETTINEN: I still want to add a little regarding lead, because there seems to be some misunderstanding. I tried to compare air pollution of lead against aqueous pollution, and came to the conclusion that the former is more important to mankind, because of its more wide nature. While water pollution usually remains rather local, air lead is not taken only by inhalation. but the air lead has polluted all aerial fallout. Our street dust is about half to two percent lead, and from this dust there are various routes to the food intake. And, therefore, we get part of our lead from the air pollution.

DR. BUCKLEY: I think this is perhaps more a comment than a question that will both save time and perhaps be in order. I would like to point out that the chemists, the analysts, have pointed out the great difficulties that they have in solving many of these problems. And I am impressed with what an extraordinary job in sensitivity they have done. I went through some arithmetic in relation to dieldrin when I heard Dr. Junk point out that 15 nanagrams per liter was the amount he was finding. And if one takes the total population of the United States, roughly 210 million, and allows them two liters of water per day, for 365 days a year, the contaminates in the water supply of the United States, that total amount of intake, at the level of 15 nanograms, is roughly five pounds of dieldrin. It is two and one quarter kilograms.

So you see, we are really talking about an extraordinarily small amount of things which our analytical chemists are able to deal with. Take that, then, and put it in terms of toxicology, and you begin to see the problems that the toxicologist has in getting hold of enough of this stuff, whatever it may be, in order to work with, to begin to have some understanding. And so, our attempts in EPA in a way are kind of a very frustrating exercise in dealing with this particular problem. If we identify these many compounds, which ones are of concern? And yet, if we don't identify them, we are not able to obtain enough of them to do basically any toxicology. We are really in a quandry!

The second point is quite different, and it has been made in other ways. But I would like to point out that in many ways, I feel as though we are in a tower of babble. We sit here using the language of the conference, English, and by no means do I mean to imply that the quality of the English is less than impeccable. It is simply that we seem to be addressing an assortment of different topics, which are closely enough related that, as we address them from our individual points of view, we don't quite draw the distinction as to whether it is sea water or drinking water, or waste water or hard water, or soft water or whatever water. And somehow or other, we need to come to grips with this.

The last point I would like to make is, it is clear if we had a different conference, and a different setting, with a different group of people, we might conclude that first things come first. We might not have the same set of first things. In this country, and in those that are largely represented here, the problems of disease, water borne disease, have been dealt with most effectively for quite a long period of time. Now we are not sure whether we have a residual problem in drinking water, in dealing with other substances present. And it is that part of it that we are groping with.

Were we in other parts of the world, it is clear to me that we would be dealing with how does one control the bacterial and other substances which clearly are detrimental.

DR. MRAK: I would like to ask Dr. Junk, or perhaps someone from EPA, should we find that chloroform should be cleaned up and taken out of the water, then we can't chlorinate any more, where do we go?

DR. JUNK: I think it is a matter, at this point in time, of comparative risks, if you will. It is clear to me that we have no adequate substitute for residual disinfection in water supply systems, and - I said no residual disinfection. Given that we have no real choice at this point in time, there are many ramifications that flow from this. I think the tradition in the water supply community has firmly established that it would be useful and desirable, and perhaps, absolutely essential to maintain a residual disinfection throughout the distribution system of water.

If people were error-free, no one ever connected the wrong pipes together, all those kinds of things, perhaps we would not need the residual disinfection. Until human nature changes, or unless we discover that this is a myth rather than a well-founded belief on the need for residual disinfection, I submit that we won't deal with the problem by anything other than chlorination.

DR. KORTE: Our knowledge on the occurrence of trace elements or inorganic elements generally in environmental samples, is normally based on two methods: neutron activation analysis and atomic absorption. Both methods only show the total element concentration. Would you find it necessary to determine for all these methods the chemical form in which they might be present in tissues, Dr. Miettinen? This would be a big issue.

DR. MIETTINEN: It is difficult to answer. It is quite evident that for some metals it is necessary, arsenic and mercury are examples. Cadmium is usually bivalent in the food and in water. I think just to determine the element is enough. But chromium is a different case again. So that it goes individually by element by element. In some cases, it is necessary to determine the chemical form, in other cases, it is perhaps not so important.

DR. COULSTON: But we have not answered the question. Can you tell which form the metal is in, its complex with organic material, how do you do this? In tissues or in any kind of organic matter? In the lead studies that I have been involved with, there is no way at present to determine how much of the lead on the street outside is in an organic form. We have no way to measure this at all analytically. We can only measure it in terms of the metal. Consequently, what we measure is the metal lead, as we do in atomic absorption, so this is a very crucial question he is raising.

DR. EGAN: This problem is not peculiar to contamination. I think, for example, it also occurs in nutrition in regard to iron and the form which the iron is present. But once again, I agree with Professor Miettinen, there is no single standard answer to the question of all metals at all times. And we have to turn to the toxicologists or the environmental biologists for guidance as to what is required. We can help, possibly devise methods for distinguishing between lead, which is bound in organic form, perhaps, or complexed with organic molecules, and that which is ionized. We already had to do this for mercury, and in slightly different context.

We have tried to help to see whether it is necessary to do organically bound cadmium or not, and, generally speaking, we think not. But there is no one standard answer for all the metals.

DR. JUNK: I am out of my field, but will take what I have heard from other scientists. I think a whole body of data on the toxicity of copper to aquatic life was in effect thrown out because the experiments were done under controled situations using a particular water. It turned out that when the particular sea water was used, or the impounded water supply, you had an entirely different set of toxicity results due to the complexing of that copper.

Now analytically, when it gets down to an analytical technique for measuring, how is this particular element tied up in the aquatic system? You have a very difficult analytical problem. To my knowledge, it is handled by those people who are concerned about it, by attempts to measure the stability constants of the known organometallic complexes.

DR. SMEETS: I want to thank you all for the very interesting discussion and the questions and comments we have had. I want to particularly thank the three panel members for the way they have answered the different problems. I think it was a very useful morning, and I, as Chairman, am very happy that we have had this occasion for discussion.

A CONSERVATIONIST'S VIEW OF DRINKING WATER

Thomas L. Kimball

On behalf of the Federation, I would like to reiterate our welcome to you as you enter the second half of this three day conference and to commend you for your attention to a subject which is of such vital concern to the health and welfare of the people of this nation.

The upcoming year represents for the Federation a year of personal as well as national commemoration. Next February, we will mark the National Wildlife Federation's fortieth anniversay as an organization which has sought to inform the public about the need for careful conservation of our natural resources, to ensure their continued benefit, not only to this generation, but also to future ones.

Of course, throughout the past forty years, the conservation of wildlife and wildlife habitat has been the special concern of the Federation. Because of this, we have devoted particular attention to the plight of the American bald eagle. Its majestic beauty symbolizes well the richness of this nation's resources while its status as an endangered species symbolizes clearly the hazards to which those resources have been subjected by decades of poorly planned industrial growth and development.

But if I had to pick out one other threatened resource which equally well symbolizes the tragedy of America's development, I would have to choose our drinking water. Who of us who has traveled abroad and contemplated the differences among nations has not been reminded of the confidence we Americans once placed in the safety of our drinking water? Certainly, in this century, drinking water in the United States has stood for the very best of what American technological know-how can achieve for the benefit of all.

Today, however, our drinking water's reputation is as threatened as the eagle. Certainly, it is one in which the public is justifiably hesitant to place its confidence after the series of private and governmental reports of drinking water contamination, not only by disease-bearing organisms, but also toxic industrial pollutants and carcinogens.

With the number of deaths due to cancer increasing at a time when researchers of the National Cancer Institute believe that between 60 and 90 percent of all human cancer is environmentally induced, the reports of carcinogens in the water we drink have been especially disturbing. Yet, the reputation of U.S.A. drinking water suffers from more than just these reports. The fact that we are not now able to establish the safety or danger of such substances heightens the public's anxiety.

For that reason, I think our drinking water also identifies an increasingly important issue of environmental protection - who should bear the burden of proof for the safety of a pollutant in the environment, when no level of safety can be set confidently? Should it rest with the advocates of a substance's use and introduction into the environment, or should it be left to the public, which must endure exposure to such pollutants, knowing that their full environmental and health effects may not manifest themselves for years, decades, and even generations?

The issue of the burden of proof is a volatile one, whose resolution in national policy will have significant ramifications scientifically, industrially, environmentally, and politically. How much evidence do we require before we take action against a man-made substance into the environment? What standards do we use in evaluating that evidence? What precautions and restrictions do we set up to protect the public health and the environment?

The public's concern for its health - particularly for its well-being from cancer - was best demonstrated last winter in the 93rd Congress' 11th Hour enactment of the Safe Drinking Water Act, shortly after the Environmental Protection Agency reported the presence of carcinogens in the New Orleans drinking water supply and the Environmental Defense Fund, Inc. statistically associated carcinogens in drinking water supplies with the incidence of human cancer along the Mississippi River.

Figuring out why Congress acts the way it does can often be risky business, but few would disagree that Congress' speedy action on drinking water legislation, which had been pending before it for more than three years, was motivated primarily by its recognition of the public's deep concern about exposure to carcinogens in the environment. That concern is one which is shared by many conservation and environmental organizations, particularly in light of the 1970 Ad Hoc Committee Report to the Surgeon General, which concluded in part: "It is essential to recognize that no level of exposure to a carcinogenic substance, however low it may be, can be at a 'safe level' for man."

Because of statements such as this and others by authorities of public stature, a growing number of environmental organizations are calling for a clearly stated federal policy on environmental cancer in particular and on toxic substances in general. Since 1971, the National Wildlife Federation, at the invitation of Congress, has testified on several occasions in support of the establishment of a federal program for regulating chemical substances before they are introduced into the environment.

Most recently, at the 39th Annual meeting of the Federation in March, the elected representatives of our affiliate members adopted a resolution urgently recommending as national policy that the manufacturers, users, or disposers of chemical substances which pose an unreasonable environmental risk should assume the burden of proof for its safety. The individuals who voted on that resolution are by no means the spokesmen of a small radical fringe. They represent outdoor sports, conservation, and environmental organizations in all fifty states and three territories - organizations which serve over a million men and women who have a broad and diverse range of interests in nature and the environment.

I think that this resolution, coming on top of NWF resolutions passed in previous years on pesticide control, air pollution, and water quality, demonstrates clearly that people are questioning and challenging how decisions are made about environmental and public health protection. I believe they are becoming irritated with arguments which illogically attempt to transfer principles of law to environmental protection. The idea that toxic substances in the environment should be considered "innocent" until "proven guilty" of health or environmental risks has little appeal to the public whose own health and well-being will tally up the evidence.

In conclusion, I would like to emphasize that the issue of the burden of proof is one which will increasingly dominate the environmental health debates - not just over drinking water regulation, but also water pollution control in general, air pollution control, pesticides regulation, solid waste management, and noise pollution control. It is an issue to which all of us - environmentalists, scientists, manufacturers, governmental officials and private citizens - must address ourselves.

CURRENT WATER QUALITY ADVANCES

H.G.S. van Raalte

When looking at your programme, only a few weeks ago, I was a little puzzled by the diversities of speakers and disciplines and the great varieties of subjects for this Water Quality Forum. Clearly there is a great multitude of aspects. Unfortunately, when preparing this talk I had not yet had the privilege of hearing Professor Korte tell us: "What is Water." Another question of course is: Which is water? Oceans, lakes, bays, lagoons, rivers, creeks, canals, drinking water or even bottled water?

The quality of water, just like the quality of life or of anything else, just is not a single parameter that can be expressed numerically. Therefore, it is not easy to discuss the quality of water scientifically. Yet, almost by instinct, we know what is meant by the quality of water. Because we are alive. We are part of water and water is part of us. About 70% by weight, to be more or less exact. It is a coincidence that the same percentage, 70% of the surface of the earth is covered by water.

All animal life is assumed to have come from the ocean and there are those who believe that the ocean still is basic to all life, including human life. Is it, then, an ominous fact to establish that in our time the surface water of the earth does not increase significantly while the world population multiplies foolishly and the standard of living should require the use of even more water per head of population?

In an obituary of Sir Julian Huxley (1887-1975), Nicolas Polunin wrote: "He had a deep feeling for Nature on the one hand and for Humanity on the other and he worried about the collision course on which they seem inexoribly bent."

I believe we all share this deep feeling of Sir Julian's, but we may not necessarily subscribe to the term "inexoribly" or even to the word "inevitably." If no one does anything to adjust the course, this collision might or might not occur. However, we humans represent the thinking part of the biosphere. As a species, we are surviving until tomorrow. We certainly have changed our environment. We can do that again and it appears we are already bending back Humanity's collision course with Nature. There is no doubt that, here and there, the quality of water has improved.

We would all agree that as regards the quality of water, the sensory threshholds of sight, smell and taste for whatever is <u>in</u> the water should not be exceeded. We must rely on analytical instrumentation. In doing that we find that the crate in which the GLC was delivered to the laboratory has indeed been a box of Pandora. Out came evidence of many nasty things we hardly even knew existed. The impact of analysis on virtually all aspects of water pollution has been enormous. There has been, in the first place, a great impact on research in water pollution. This, in its turn, has certainly led to advances in water quality, and these have been five-fold:

Advances in Water Quality:

through

1. Advances in Knowledge and Understanding
2. Changes in Attitude
3. Advances in Technology
4. Advances in Legislation and Regulation
5. Education of the Public
   (including) Industry

Based on advances in analysis, there were first advances in knowledge and understanding of the nature of pollutants, natural and man-made, industrial and social, chemical and microbiological.

This was followed by advances in knowledge and understanding of the effects of the pollutants on living things in the aquatic environment.

This knowledge and understanding caused a change in attitude in many industrial managements. It was realized that there is a limit to the capacity of the "Commons." In using the term "Commons," I do not mean the lower chamber of the British Parliament, although there may be a limit to their capacity, too. The term "Commons" as used here is taken from Hardin's worth-reading article: "The tragedy of the Commons" and it means the common municipal grazing ground, owned by the entire municipality where every herdsman from the community could graze his animals. In using the word "Tragedy," Garrett Hardin cited the philosopher, Whitehead: "The essence of dramatic tragedy.....resides in the solemnity of the remorseless working of things."

"It did not much matter," Hardin said, "how a lonely American Frontiersman disposed of his waste." Hardin's grandfather used to say: "Flowing water purifies itself every ten miles," and thus, Hardin continued, "Using the Commons as a cesspool does not harm the general public under frontier conditions, because there is no public; the same behaviour in a metropolis is unbearable."

Now, everyone realizes that the Commons, as a place of waste disposal, has to be abandoned. As regards this change in attitude, here is an example of how advance in water quality can be achieved by appropriate action, once the connection of things is understood.

In 1965/1966, it was reported that the numbers of sandwich terns on a small island bird sanctuary off the coast of the Netherlands was decreasing alarmingly. In the birds, a mixture of residues of some organochlorine insecticides was found, amongst which unusually high residues of telodrin and endrin. The pattern of residues of the various OC's in the fat and organs of the birds was inconsistent with the use-pattern and the biological half-life of the compounds. Moreover, endrin and telodrin were not used at all anywhere near the island sanctuary and endrin particularly, has a remarkably short half-life in animals. Neither, for that matter, was there any important agricultural use of the other OC's in the vicinity.

After a thorough investigation in cooperation with biologists and toxicologists from Utrecht University (Koeman), the problem was unraveled and on its way to being solved.

In the bloodstream of a large insecticide manufacturing plant, the precursors of endrin and telodrin escaped into the water of the second petroleum harbour, and as all chlorinated cyclodienes, rapidly sorbed onto mud particles on the bottom, and were, initially, not detected in the water of the harbour. The sorption onto mud particles is so firm that leaching and displacement is insignificant. However, the harbour is dredged regularly to insure its continued accessibility. The dredger deposits the mud on barges. These are towed to open sea and the mud is dumped, but for some time, remains in suspension. Some of it is caught in the tide and carried northwards toward and along the coast. Some of the insecticide from the still suspended mud particles ends up in small crustacea and marina beachworms, which are eaten by the terns, after having been converted to telodrin and endrin, respectively.

There have been advances in technique which have enabled a considerable reduction in pollution of the water and a new technology will be developed which will enable production without creating the large amounts of waste that so often still accompany current industrial activities.

The fourth source of advances in water quality is new legislation and regulation. Hardin's "Tragedy of the Commons" clearly shows that legislation is a "must."

"We are still struggling," said Hardin, "to close the Commons to pollution by automobiles, factories, insecticide-sprayers, etc. In a more embryonic state is our recognition of the evils of the Commons in matters of pleasure. Every new enclosure of the Commons involves the infringement of somebody's personal liberty." Thus, we will have to accept that we cannot close the Commons only to the other herdman's animals.

Legislation is nothing new, however. We see, as an example, the Netherlands, probably the only country in the world whose historical records begin before most of the land existed. In Roman times, Tacitus tells of a "large number of lakes north of the River Rhine, fringed by a chain of sand dunes." With the flooding that continued until the 16th century, Holland was more sea than land. Then, along came an engineer called Jan Adriaanszoon Leeghwater (which means "empty water" as well as "low tide"). With a large system of windmills, dikes, and polders, he created land from sea and lakes. Holland is still reclaiming land and farming from scratch. In such a country, legislation of the water management is paramount.

The country is small, only 40.844 km^2, or about 15,765 square miles, of which 20% is water. It is densely populated by 13.5 million people, 385 per sq.km (EEC:164). Two-fifths of the country is below sea level.

Much of the waterways are shallow and unnavigable except for small craft. The Dutch waterborne transport is heavy. The regulations distinguish between primarily commercial waterways, primarily recreational and those for both commerce and pleasure. The legislators are making every effort to harmonize the interests of commercial navigation with the various forms of recreation and nature conservation. In addition, today 40% of all the drinking water comes from surface water and it is estimated that in the year 2000, this percentage will be 60.

Several institutes, agencies and organizations cooperate in carrying out the various regulations:

R.I.D.: National Institute for the Provision of Drinking Water: establish specifications for drinking water.

R.I.V.: National Institute of Health: check the quality.

K.I.W.A.: Institute for testing of Waterworks Articles: establish and check requirements for articles and materials.

M.W.: Ministry of Waterways and Transport: regulate distribution of surface water, lakes, canals, etc. They issue approval for waste-disposal in waterways.

Waterschappen: Polder authorities who look after the interests of their inhabitants, agriculture, dike maintenance, etc.

Internationally, the Convention of Oslo regulates dumping of materials in the sea (February 2, 1972). This agreement is extended within the jurisdiction of the European Communities by their EEC Council directive: "Prevention of Marine Pollution from Land-based Sources," to the estuaries and rivers.

The fifth source of advance in water quality is education of the public. This includes legislators and Industry. Just like legislation itself, this education should be practicable, otherwise it may become counterproductive. It should not be based on speculation and fear mongering; it should be based on scientific facts.

The myth, for instance, of the immutable irreversibility of "dead water," is not a scientific fact. This legend is put to the pillory by numerous observations on advances in water quality. I shall give you four examples:

1. Salmon have recently been seen again in the River Rhine.
2. The green frog, _rana_ _esculenta_ _L_., is back in the Dutch polders and recently even made its debut on Dutch television.
3. In 1957-1959, a survey was undertaken on the occurrence of freshwater fishes in the River Thames, the longest river in the British Isles, and particularly on fishlife in the lower Thames in the London area. There was no evidence that any fish were living in the Thames below central London. Preventive measures were then taken to reduce pollution from both domestic and industrial wastes. In and around 1962, fish were observed farther downstream into London than they had been found five years earlier. Since then, fresh water fishes have gradually extended their range down the river. The conclusion was that the Thames fish fauna is re-establishing itself very quickly.
4. In the Western Hemisphere, a similar experience has been encountered on the island of Curacao. In and around a large bay in open connection with the Caribbean sea, one of the largest oil refineries in the world is situated. When the refinery and tanker shipments were smaller, back in 1937, and not so many people living there, I have often seen fish being caught and people swimming, so to say in between the refinery installations and the oiltanks. Over the years, with the extension of the refinery and the shipping and the increase in population, the water in the bay became "dead water." There was no evidence of fishlife.
   The installation of oil-interceptors, oil-skimmers, the improvement of the draining system, sewage treatment and disposal, the dissolved oxygen content of the baywater increased considerably within five years' time, and fish are multiplying again, like they did at the wedding in Kana.

THE SCIENTIFIC BASIS FOR THE ESTABLISHMENT OF WATER QUALITY STANDARDS IN SWITZERLAND

for Hans Gysin

by Robert F. Curran

FROM THE GUIDELINES TO A DECREE ON WATER AND EFFLUENT QUALITY STANDARDS

In a decree issued on September 1, 1966, the Swiss Federal Government describes guidelines for the quality of effluent to be discharged. In this document two different sets of data are mentioned:

- one set for effluent which is discharged directly into rivers or lakes
- another set which describes the conditions under which effluent is to be discharged into municipal (or private) sewer systems which are connected to a mechanical-biological treatment plant.

The figures represent values which apply to dry weather conditions and the established goals should not be reached by dilution of effluent for example with cooling water. Where possible, effluent from industry and municipal effluent should be treated jointly in a biological system.

In addition, the decree describes general and special quality requirements for the effluent to be discharged into rivers and municipal sewer systems respectively.

The guidelines of the mentioned decree are in force up to the present day, however, since their issue in 1966 additional legislation was published. In a vote of June 6, 1971 an addition of the Federal Constitution was accepted by the Swiss population with an overwhelming majority which asks in general terms for the protection of man and his natural environment against toxic and noxious influences. This amendment to the constitution is the legal basis for the general environment protection law which is in preparation and which will cover air and noise specifically. The law to protect rivers and lakes was adopted by the Federal Government in June 1971 and accepted by the Federal Assembly in October 1971 already. This law, which is the basis for all water protection measures, is in force since July 1, 1972. The Federal Government is responsible for the supervision of the execution of the water protection law while the enforcement is left to the cantons and communities respectively. The law allows a ten-year clean-up time which means that by mid-1982 all mechanical-biological effluent treatment plants have to be fully operational.

WATER CONTROL STANDARDS

Based on guidelines on effluent quality of 1966, an interparlamentary commission prepared a draft for a decree with regulations to safeguard or improve the quality of water bodies. It describes in a general manner the conditions which have to be reached in the rivers and specifies standards at the discharge points into running and stagnant water. It specifies some requirements as far as the influence of treated effluent on the fauna and flora of rivers is concerned. The production of drinking water from rivers into which treated effluent is discharged has to be guaranteed as well as hygenically acceptable conditions for bathing in these rivers. In addition to the more general claims, a list of standards for inorganic and organic compounds as well as some "summation parameters" such as COD, TOC, etc. are given. The document is ready for signature by the Federal Government and should be in force soon thereafter.

The parameters of the new legislation are divided into three categories:

1. Quality goals for running and stagnant water bodies,
2. Requirements for the quality of effluent at the discharge point into rivers or lakes,
3. Requirements for the quality of effluent when discharged into a municipal sewer system.

It has to be stressed that the parameters in categories 2 and 3 are mostly given as well defined figures while for the water quality of rivers and lakes goals are set only which should be reached when the requirements prescribed in the other categories will be fulfilled.

During the earlier phase of the construction of integrated mechanical-biological effluent treatment plants, mostly municipal works were constructed whereby the local industries gave their effluents, sometimes pretreated, sometimes not, to these municipal plants. In contrast, the major Swiss chemical industries built a number of joint municipal/industry plants over the last five years after having done preparatory studies for several years before that period. Basically two types of joint operations exist, namely a joint treatment of the industrial and municipal waste water at the same site. In both cases, the sludge-handling, drying, eventually incineration and deposition is done jointly.

GENERAL THRESHOLD PARAMETERS

TEMPERATURE

A maximal temperature rise of $3^{o}C$ in the receiving water body is envisaged whereby a temperature of 25°C should not be surpassed. When treated effluent is discharged into rivers and lakes, its temperature at the receiving site is not allowed to be higher than 30°C. The temperature of effluent discharged into a municipal sewer system is limited to 60°C at the discharge point whereby the temperature in the sewer system should not rise above 40°C. The discharge regulations may be adjusted to the local conditions on a case-to-case basis. This is a typical example where parameters were set arbitrarily based on some experience rather than drawn from scientific data.

COLOR

A rather poor definition is given as far as color is concerned. There the law says that - as a consequence of effluent discharge - no discoloration should occur in the river and colored effluent is only allowed to be discharged into municipal sewage systems providing that discoloration takes place in the biological treatment plant. Experience shows that in neither case this is true so that discoloration has to be achieved at the plant site. It is highly questionable by the way whether a more or less colorless effluent from a dyestuff manufacturer is less harmful for the river biology than slightly colored effluent still containing unchanged dyes.

TOXICITY

For this parameter at least the criteria are based on defined tests for the effluent quality. For the rivers, however, "no toxicity" is envisaged which is a rather unprecise goal. For the discharged effluent an exact test procedure is given which has to be executed based on the conditions of the respective water body in a zero to five-fold dilution. Thereby the fish should show no toxicity symptoms within 24 hours exposure. The size, age and the number of test fishes are defined in a federal decree so that this parameter can be regarded as based on scientific facts. The only questionable part that is arbitrary is the variable dilution factor of the receiving water body. A much less precise requirement is given for the effluent discharged into municipal system. The law specifies there that the quality of the effluent must be such that no biological reactions should be inhibited and that neither the sludge quality nor the slude treatment procedure should be negatively influenced due to toxic materials.

SALT CONTENT

The salt content of discharged effluent into rivers should not influence in any way surface or sub-surface water. Effluent introduced into sewer systems should not negatively influence effluent treatment plants.

TOTAL UNDISSOLVED MATERIALS

Effluent discharge should not cause sludge formation in rivers, a goal which is not very clear. However, the standard for effluent to be discharged into rivers which is fixed at 20 mg per liter on a 0.45 mm filter in four out of five samples of a 24-hour average is clear. If effluent is discharged to municipal systems, the canton can fix the conditions; this is not a very precise recommendation at all. Here again we have rather arbitrary requirements.

SETTLING SOLIDS

As a goal no sludge formation in the rivers is envisaged. As far as effluent quality is concerned, a maximum of 0.3 ml per liter after settling time of two hours in an Imhoff cone is allowed.

pH

As a consequence of an effluent discharge the natural pH of the river should not be influenced. At the site of the effluent discharge a pH of 6.5 to 8.5 is allowed. If the waterflow is sufficient, the pH may be as high as 9. In order to omit an undesired increase of salt formation in the river it is tolerated in certain cases to make use of the acid binding capacity of a river for neutralization. These values were most likely taken from legislation of other countries.

THRESHOLD VALUES FOR INORGANIC COMPOUNDS

For a number of inorganic compounds, especially for metals, concentrations are specified which are quality goals for rivers and lakes. The values given for effluent quality standards are for most metals identical at the points of discharge into rivers and into sewer systems respectively. The goals envisaged for rivers are lowest for mercury and cadmium, being 0.001 mg/l in the first and 0.005 mg/l in the second case. While in the case of mercury 0.01 mg/l is tolerated at the effluent discharge point which is ten times the envisaged concentration in the river, the analogue figure for cadmium is 0.1 mg/l which is only twenty times the concentration in the effluent. In the case of copper this ratio is even more extreme as 0.01 mg/l is accepted in the river while at the discharge point 0.5 mg/l is allowed which is equivalent to a ratio of 1:50. For these metals the dilution factors from effluent to rivers vary between 10 and 50, while for the toxicity tests dilutions of 1-5 are granted only. This does not seem very logical.

In comparison with the guidelines of 1966, it is remarkable that in the case of cadmium the tolerated concentration was decreased in the river by a factor of 200, at the discharge point by a factor of 10. The respective factors for mercury are 100 and 10.

Accidents and incidents such as Minamata or Itai-Itai may have been the main motivation for making the legislation for these two metals more strict. In the case of mercury new findings namely the conversion of inorganic mercury salts into organo-mercury compounds, especially into methylmercury derivatives by micro-organisms may form a scientific basis for the change in legislation.

While all the standards discussed so far are relative concentration values, EPA in this country started to limit effluent loads. Ciba-Geigy US was asked in 1972 to reduce the daily discharge of the following metals to one pound per day each: zinc, copper, chromium, lead, cobalt, nickel, irrespective of the production volume. As this plant in New Jersey is an important producer of metal complex dyestuffs, a reduction to the indicated levels proved to be impossible with the present technology. As there was a suspicion that these standards had been fixed by EPA more or less arbitrarily, our group company made - after having studied the technical realization of the problem very carefully - a counterproposal for the establishment of higher standards for these metals.

Under US conditions it is a normal procedure to challenge such proposed values and to submit to EPA proposals which can be reached with economically feasible technical means. Under Swiss conditions it would be practically impossible to challenge any standards which are published in a decree. The meaning of the parameters published in the two countries for industry is quite different. In our country these values are carved into granite while in your country the legal remedy of an adjudicatory hearing is available to gain relief from an unreasonable permit limitation.

Let us now have a look at some parameters issued in other countries. The following table gives the effluent quality standards to receiving water for a number of metals:

	FRG	DK	IT	CH	US	J	USSR
Cd	1.0	0	0.05	0.1	0.05	0.1	0.1
Hg		0	0.005	0.01		below analyt. sensitivity	
Co	0.2	0.5	1.0	0.5	0.5	3.0	0.5
Cr	1.0	0.2	2.05	2.1	1.5	2.0	1.1
Zn	5.0	1.0		2.0	1.0	5.0	1.0

(all figures are mg/1)

As you may take from this table, the congruence of the standards is not very good and even within the common market area there is not any single parameter which is identical for all three mentioned countries.

Let us turn back to the Swiss water protection law, and discuss

THRESHOLD VALUES OF ORGANIC SUMMATION PARAMETERS

DISSOLVED ORGANIC CARBON (DOC)

The goal to be reached in rivers and lakes is 2 mg/l for DOC expressed as C after filtration of an effluent sample through a membrane filter of 0.45 mμ. For the effluent quality standards, two cases are differentiated whereby case 1 is applicable for most industry treatment plants. For this case the upper limit for dissolved organic carbon is 10 mg C/l. This value is based on an inlet concentration of total organic carbon (TOC) of less than 65 mg C/l. For higher TOC inlet concentrations higher dissolved organic carbon levels are admissible providing that at least 85% purification is reached according to the following equation.

$$100 \cdot \left(1 - \frac{\text{mg DOC in the purified effluent}}{\text{mg TOC in the discharged effluent}}\right)$$

Case 2 describes the conditions for small - mostly municipal - effluent treatment plants with rather uniform effluent compositions with minimal amounts of carbon derivatives that could have undesirable effects on the receiving water body. For inlet concentrations higher than 65 mg C/l higher DOC limits are allowed providing that the purification effect is according to the equation above at least 75%.

TOTAL ORGANIC CARBON (TOC)

Swiss legislation does not specify any goal for TOC to be reached in running and stagnant water bodies. For effluent the TOC parameter is limited to 7 mg/l above the admissible DOC level. This means that 17 mg/l of TOC is allowed for the above mentioned case 1 and effluent having a DOC-concentration lower than 65 mg C/l. For higher inlet concentrations of DOC the cantonal authorities can allow higher TOC limits on a case-to-case basis.

The discussions on the fixation for standards for DOC and TOC between the representatives of universities and industry were long and hard, but the above mentioned compromise was finally reached in the preparatory commission. The elaboration of these standards is a typical example of the Swiss way to reach a common goal: good political sense rather than highly scientific methods will be applied to reach a common goal in an economic way.

BIOCHEMICAL OXYGEN DEMAND (BOD)

The biochemical oxygen demand in rivers is not allowed to rise above 4 mg O_2/l when effluent is discharged. The BOD for effluent itself is limited at the discharge point to 20 mg O_2/l (for the judgment an average concentration over 24 hours has to be considered, whereby the prescribed value has to be reached in at least four out of five cases). Furthermore, the effect on BOD-reduction, based on a settled effluent, has to be at least 85%. The cantons are entitled to allow in coordination with the Federal Office for Environment Control BOD values higher than 20 mg/l especially if this leads to a reduction of the total effluent load and does not endanger the water body.

LIMITING VALUES FOR ORGANIC COMPOUNDS

For a number of organic chemicals specific limiting values are defined in the new legislation. Aromatic amines, hydrolyzable fats or oils, total hydrocarbons, chlorinated solvents, lipophilic low volatile organochlorine compounds, organochlorine pesticides and phenols are specifically mentioned. The allowed concentrations for the effluent quality standards are in general more than ten-fold higher than the envisaged concentration in the receiving water. Sometimes the cantons are permitted to fix the respective values in consultation with the Federal Environment Control Authorities.

An attempt to make a comparison between some organic summation parameters in a number of countries failed as most of these standards were fixed on a regional rather than on a country basis. For BOD, Belgium has fixed different standards for navigable and non-navigable rivers, Japan has different standards in different prefectures and Denmark has one value for the entire country.

CRITICAL EVALUATION OF THE NEW WATER QUALITY STANDARDS

After having discussed the standards which will become the new law for effluent quality in Switzerland, and after having compared some of these data with the respective limits in other countries, an evaluation of the Swiss standards leads to the following result:

Although some of the parameters are based on scientific facts, most others are taken from experience or are the result of negotiations between the authorities and interested circles or stem from existing legislation of other countries. Contrary to other countries there is in principle only one set of standards for all rivers. As practically all Swiss rivers may be suppliers for drinking water, this uniformity is a logical consequence. It is, however, possible that under certain circumstances the cantons can fix more severe standards for highly populated or highly industrialized areas. In comparison with other countries in Western Europe, the Swiss standards are somewhere in the middle, while especially the Scandinavian legislation is in many respects more rigid, the common market countries with the exception of Denmark, generally have less severe laws. The standards vary considerably from one country to another, which is due to the fact that many effluent standards are the result of negotiations rather than being based on scientific evaluations. This is especially disturbing for international rivers such as the Rhine where no common standards exist for a great many inorganic or organic chemicals. A serious attempt, therefore, should be made to standardize the parameters in such a way that the respective industries in these countries are not discriminated. Although international organizations like the Rhine Protection Commission are advocating this harmonization, not much progress has been made so far. A standardization of quality parameters is urgently needed as in some countries new legislation is already in preparation and the discrepancies of present legislation will then become still more pronounced.

FUTURE LEGISLATION

There is no doubt whatsoever that a complete fulfillment of the standards set by the Swiss legislation will be a time and money consuming process. In addition, it will force industry to apply new and improved technology for many chemicals needed to safeguard human welfare. This has to be kept in mind before any further steps are undertaken. Among environmentalists new and more severe legislation is being constantly advocated. As existing standards were fixed partly arbitrarily, partly by compromise, and a few based on scientific data only, there is no harmonization between the countries at the present.

No new and more stringent standards should be issued before

1. the improvements due to present legislation are known,
2. a need for new legislation exists,
3. a sound scientific basis for new standards is established.
4. a harmonization at least between the industrialized countries can be reached.

We cannot continue to just cut present standards into half every second year without carefully investigating which benefits can be reached for the environment and what economic impact the introduction of new standards may have. At the beginning of the 1980's, in our country, 90% or more of the biodegradable organic carbon will be removed with integrated mechanical-biological effluent treatment plants. It is possible that improved technology will even lead to a higher purification so that biodegradable compounds should not be a problem after the early 1980's. In order to clean up the rivers and lakes respectively from non-degradable organic materials and from inorganic salts, the application of specific methods will become necessary. Whether these additional purification methods are based on chemistry, physics or a combination of the two, will depend on the specific problems. Whenever possible, the decrease of non-biodegradable materials will have to be done at the plant site where specific methods can be applied as long as the effluent is more uniform and concentrated. For the purification of the rivers from inorganic salts, in many cases recovery systems are the only solution. Sulfuric acid, for example, one of the big polluters of rivers due to chemical processes, has to be concentrated in such a way that it can be worked up into "new" acid which goes back to the process again. In addition to the attempts to decrease both TOC and salts, improved technology has to be worked out. Products which may be specifically objectionable for rivers may have to be substituted by less harmful chemicals, which means that in certain cases not only processes but also products may have to be changed.

Can industry live with the present limits or possibly with more rigid standards?

In order to reach the standards of the new legislation, the chemical industry has to spend over a period of years about 10% of their total investment. Additional money and energy will be required to run the treatment plants and to fulfill the new legislation. It is our belief that in general, especially in view of some exceptions which may be granted, industry will be able to fulfill the standards by 1982. Every effort should be undertaken to reach by that time a harmonization of standards in Western Europe and hopefully in all industrialized countries. Furthermore, industry wishes that any stiffening of the present legislation at a later date should be strictly based on scientific data and enough time should be provided when new parameters are introduced. A very careful evaluation should be made about the economic implications of any future legislation, and new and more strict laws should be issued only when the progress thus to be reached justifies the extra expenses. All those who ask for tougher pollution control legislation have to keep in mind that they, as end-consumers, finally have to pay for the degree of clean-up they are asking for.

INTERNATIONAL STANDARDS, NORMS AND CRITERIA FOR WATER QUALITY IN EASTERN AND WESTERN EUROPE

Michael J. Suess

Resume of Activities of the Unit for Control of Environmental Pollution and Hazards

Progress achieved

Within the overall framework of the Promotion of Environmental Health (PEH) Service, the Unit for Control of Environmental Pollution and Hazards is concentrating at present on intercountry activities in the sectors of water and air pollution control, solid waste management and radiation protection, all of which are part of the long-term programme in environmental pollution control adopted by the Regional Committee in 1969. The principal mode of action for the implementation of the programme is the production of a series of manuals which will provide information and recommendations for pollution control and health protection programmes and to serve as aids to the decision-making and management systems of governmental administrations and institutions concerned with the quality of the environment. Also, the manuals will provide national scientific specialists and engineers with useful information based on international practice. While originally planned for use in the European Region, the manuals should prove equally useful to governments and scientific communities in other regions. Four manuals are now under preparation:

- Analysis for Water Pollution Control,
- Urban Air Quality Management,
- Solid Waste Management,
- Non-ionizing Radiation Protection.

A fifth manual, on Air Pollutants from Industrial Sources, is under way.

Ten of the twelve chapters of the water pollution control manual are already completed or are in the final stage of revision following a total of eight working groups which have reviewed and discussed them. Two more chapters are now under preparation and will complete the first edition of the manual, expected to be ready for print in its entirety in 1977. The manual on solid waste management is expected to contain some two dozen chapters, of which two are complete and four more are under preparation. Lack of funds and manpower has held back the progress of this work. Two of some half dozen chapters of the manual on non-ionizing radiation are almost complete and one more is under preparation. The first draft of the manual on air pollutants from industrial sources is expected to be prepared by the German Engineering Association (VDI) for this Office, which is carrying out the project in close collaboration with the United Nations Economic Commission for Europe.

The unit is also engaged in other activities aimed at promoting intercountry collaboration and examining the validity and usefulness of the manuals, as a whole or in parts. Thus, an intercalibration project has been developed, according to which various analytical methods for water examination are considered, as well as the comparability of the results received from over 30 participating laboratories in about 20 countries. The first programme deals with intercalibrated pH, alkalinity, electrical conductance and chlorides. Four other programes, now under way, will deal with BOD, COD, TOC, heavy metals (Cd, Cu, Fe and Zn) and anionic detergents. A fifth programme will deal with five radionuclides and will be conducted in collaboration with the International Atomic Energy Agency.

To promote further cooperation among riparian countries in safeguarding water bodies, meetings were held to discuss and develop appropriate ecological and pollution control work plans which, in some cases, would adopt the manual on Analysis for Water Pollution Control for part of their work. Such meetings were held to discuss specifically the Rhine, the Danube, the North Sea and the Mediterranean (the last three by other units in PEH). As a result of the Rhine meeting, a proposed plan was developed for an ecological study of the Rhine Valley. Awareness of the serious inland eutorphication problem of surface waters has led to the preparation of an independent short monograph on the causes and effects of eutrophication and on means of controlling it.

Within the air pollution control sector, epidemiological studies have taken place in five countries on the correlation between air pollution and chronic respiratory diseases in children.

Finally, in view of the growing concern of European authorities about the many construction plans for nuclear power reactors, a series of meetings will take place to discuss and make recommendations on the implications of nuclear power production for man's health and the environment.

Future programme

Activities in progress will be continued, as is natural in a long-term programme. Thus, for example, it is anticipated that outdated manuals or certain chapters of manuals will be revised and replaced as and when necessary. At the same time, new manuals will be prepared, such as one on the survey and classification of inland surface waters. Further studies will be developed, which could also examine and verify the technical and operative relevance of the manuals. Such activities will include ecological and eutrophication studies and water quality monitoring. The preparation of recommendations for the extension of drinking water reservoirs for recreational use would be another important activity within the overall water pollution control sector.

The epidemiological studies on the effects on health of air pollution will continue and other interested countries are expected to participate. Also, attempts will be made to correlate the results of these studies with those of the Commission of the European Communities, which are executed along very similar lines. A new area of study on indoor climate (micrometeorology) will be started with an attempt to determine exposure levels and to correlate them with housing quality. The harmonization of analytical methods and instrumentation for the measurement of air pollutants and their continuous monitoring will be an important link within the overall chain of activities of the air pollution control sector.

Within the solid waste sector, investigations will be made on needs, methods and economics of waste recovery, recycling and conversion to energy. These will be related to an operational research study on the collection and transportation of waste, on the one hand, and to the problems of analyzing waste composition, on the other. A study of the effect of leaks from landfills on the quality of ground and surface waters will also be of importance.

The evaluation of the effects of nuclear power production will be continued and attention will be paid to the effects of thermal air effluents and cooling towers on meteorological conditions and, consequently, on health. Attention will also be paid to the effect of thermal water effluents on the aquatic ecosystems. With respect to non-ionizing radiation, consideration will be given to problems of threshholds and differences in biologic effects due to chronic and acute exposure, the effects of electro-magnetic fields, the hazards in the use of ultrasound applications and the dose-response relationship in the generation of skin cancer by ultra-violet radiation.

TABLE 1. List of Reports and Publications on Environmental Health and Pollution Control[1]

Ref. no.[2]	Title[3]	Place and date	Language[4]	Subject[5]
Reports				
EURO 159.4	Water Pollution Control (C) (pp. 39)	Budapest 11-15 Oct 1966	E F R	W
EURO 0383	The Public Health Implications of the Widespread use of Radioactive Materials and the Disposal of Radioactive Wastes to the Environment (WG) (pp. 40)	Copenhagen 17-19 Oct 1967	E F R	R
EURO 1143	The Health Effects of Air Pollution (Sp) (pp. 73)	Prague 6-10 Nov 1967	E F R	A
EURO 0664	European Standards for Drinking-Water, Second Revised Edition (WG) (pp. 56)	Copenhagen 18-21	E F	W
EURO 0415	Trends and Developments in Water Pollution Control in Europe (WG) (pp. 31)	Copenhagen 23-26 Sep 1969	E F R	W
EURO 3103	Trends and Developments in Air Pollution Control in Europe (WG) (pp. 46)	Copenhagen 19-22 Jan 1971	E F R	A
EURO 3119	Automatic Water Quality Monitoring (Sm) (pp. 29)	Cracow 29 Mar - 2 Apr 1971	E F R	W
EURO 3402(1)	Development of Solid Wastes Programme (WG) (pp. 33)	Bilthoven 4-6 May 1971	E F R	S
EURO 7902	Modern Trends in the Prevention of Pesticide Intoxications (C) (pp. 34)	Kiev 1-4 Jun 1971	E F R	P
EURO 3105	Accidental Pollution of Inland Waters (C) (pp. 57)	Bucharest 27 Sep - 1 Oct 1971	E F R	W
EURO 3901	Development of the Noise Control Programme (WG) (pp. 25)	The Hague 5-8 Oct 1971	E F R	N
EURO 4701	Health Effects of Ionizing and Non-Ionizing Radiation (WG) (pp. 33)	The Hague 15-17 Nov 1971	E F R	R
EURO 3114(1)	The Long-term Effects on Health of Air Pollution (WG) (pp. 47)	Copenhagen 14-18 Feb 1972	E	A
EURO 3106(1)	Manual on Air Quality Management in Europe (WG) (pp. 7)	Dusseldorf 27-29 Mar 1972	E	A

TABLE 1 (cont.)

Ref. no.[2]	Title[3]	Place and date	Language[4]	Subject[5]
EURO 3109(1)	The Hazard to Health of Persistent Substances in Water (WG) (pp. 17)	Helsinki 10-14 Apr 1972	E	W
EURO 3106(2)	Manual on Air Quality Management in Europe (WG) (pp. 12)	Frankfurt 13-15 Sep 1972	E F R	A
EURO 3402(2)	Solid Waste Management in Europe (WG) (pp. 8)	Copenhagen 11-13 Oct 1972	E F R	S
EURO 3110(1)	Analytical Methods in Water Pollution Control (WG) (pp. 9)	Copenhagen 7-10 Nov 1972	E F R	W
EURO 3128(1)	The Hazards to Health and Ecological Effects of Pollution of the North Sea (WG) (pp. 14)	Bilthoven 6-8 Dec 1972	E F	E
EURO 3114(2)	Study on Chronic Respiratory Diseases in Children in Relation to Air Pollution (WG) (pp. 6)	Rotterdam 26-28 Feb 1973	E F R	A
EURO 3114	Working Protocol and Recording Forms (pp. 24)	Copenhagen 26 Mar 1973	E	A
EURO 3006(1)	Study on Manpower Requirements in Environmental Health (SC) (pp. 45)	Copenhagen 25-29 Jun 1973	E F R	M
EURO 3110(2)	Intercalibration Programmes (WG) (pp. 6)	Copenhagen 28-30 Aug 1973	E F R	W
EURO 3604	The Public Health Aspects of Antibiotics in Feedstuffs (WG) (pp. 37)	Bremen 1-5 Oct 1973	E F R	F
EURO 3170	Health Hazards from Exposure to Microwaves (EG) (pp. 11)	Copenhagen 22-23 Oct 1973	E F R	R
EURO 3109(3)	Hazards to Health and Ecological Effects of Persistent Substances in the Environment (WG) (pp. 3)	Stockholm 29 Oct - 2 Nov 1973	E F R	E
EURO 3129	Toxicity Information Systems (PM) (pp. 7)	Stockholm 2-3 Nov 1973	E F R	I
EURO 3104(2)	Long-Term Programme in Environmental Pollution Control (PM) (pp. 8)	Copenhagen 13-14 Nov 1973	E F R	-
EURO 3404(1)	Toxic and Other Hazardous Waste (WG) (pp. 15)	West Berlin 20-23 Nov 1973	E F R	S

TABLE 1 (cont.)

Ref. no.[2]	Title[3]	Place and date	Language[4]	Subject[5]
EURO 3109(2)	Hazards to Health and Ecological Effects of PCB's in the Environment (WG) (pp. 2)	Brussels 3-7 Dec 1973	E F R	E
EURO 4108	The Health Aspects of Urban Development (Sm) (pp. 70)	Stuttgart 3-7 Dec 1973	E F R	-
EURO 3110(3)	Automated Water Monitoring and Analysis of Water (WG) (pp. 6)	Budapest 11-14 Feb 1974	E F R	W
EURO 3114(3)	Study on Chronic Respiratory Diseases in Children in Relation to Air Pollution (WG) (pp. 69)	Dusseldorf 17-19 Apr 1974	E	A
EURO 3114(4)	Physical and Chemical Examination of Water (WG) (pp. 14)	Prague 21-24 May 1974	E F R	W
EURO 3104(3)	Long-Term Programme in Environmental Pollution Control (EG) (pp. 8)	Copenhagen 10-14 Jun 1974	E F R	-
EURO 3106(3)	Pollution by Lead and other Nonferrous Metallurgical Industries (WG) (pp. 9)	Brussels 2-4 Jul 1974	E R	A
EURO 3128(3)	The Oslo Fjord Pilot Project on Studies of Sublethal Effects in Marine Organisms (SC) (pp. 2)	Gothenburg 29-30 Aug 1974	E F R	E
EURO 3006(2)	Study on Manpower Requirements in Environmental Health (SC) (pp. 2)	Copenhagen 2-6 Sep 1974	E F R	M
EURO 4905(8)	Epidemiological Surveillance of Long-term Health Effects of Environmental Hazards (WG) (pp. 17)	Copenhagen 24-26 Sep 1974	E R	-
EURO 3110(5)	Design of Measurement Systems, Sampling Programmes and Data Processing (WG) (pp. 6)	Koblenz 1-4 Oct 1974	E F R	W
EURO 3170(1)	Health Effects from Lasers (WG)	Dublin 21-24 Oct 1974		R
EURO 3104(1)	Regional Residuals Environmental Quality Management Models (WG) (pp. 10)	Rotterdam 22-25 Oct 1974	E R	M
EURO 3125(1)	Guides and Criteria for Recreational Quality of Beaches and Coastal Waters (WG) (pp. 31)	Bilthoven 28 Oct - 1 Nov 1974	E	W
EURO 3109(4)	Methods for Studying Biological Effects of Pollutants (WG) (pp. 4)	Moscow 19-22 Nov 1974	E F R	E

TABLE 1 (cont.)

Ref. no.[2]	Title[3]	Place and date	Language[4]	Subject[5]
EURO 3128(4)	Study of Sublethal Effects on Marine Organisms in the Firth of Clyde, The Oslo Fjord and the Wadden Sea, (WG) (pp. 38)	Wageningen 2-4 Dec 1974	E	E
CEP 207(1)	Ecological Aspects of Pollution in the Rhine (WG) (pp. 39)	Bilthoven 9-13 Dec 1974	E	E, W
EURO 3134	Coastal Pollution and other Environmental Health Problems in the Mediterranean (WG) (pp. 14)	Copenhagen 16-19 Dec 1974	E	W
SES 004	Meeting of Leaders of Training Courses in Human Ecology (pp. 2)	Copenhagen 14-15 Feb 1975	E F F	E, T
CEP 208	Study and Assessment of the Water Quality of the Danube (WG) (pp. 6)	Copenhagen 3-7 Mar 1975	E F R	W
CEP 206(6)	Radiological Examination of Water (WG) (pp. 12)	Vienna 21-24 Mar 1975	E	W
CEP 206(7)	Bacteriological and Virological Examination of Water (WG) (pp. 11)	Mainz 21-25 Apr 1975	E	W
CEP 303	Air Pollutants from Industrial Sources (PM) (pp. 4)	Dusseldorf 28-29 Apr 1975	E	A
CEP 206(8)	Biological Examination of Water (WG)	Brussels 17-20 Jun 1975	E	W
Publications[6]				
EURO 2631	Lang, J. and Jansen, G., *The Environmental Health Aspects of Noise Research and Noise Control*, WHO Regional Office for Europe, Copenhagen, 1970 (pp. 97)		E F R	N
	Hepple, P., Editor, *Water Pollution by Oil*, The Institute of Petroleum, London 1971 (pp. 393). (Proceedings of a seminar at Aviemore, Scotland, 4-8 May 1970, assisted by the Regional Office)		E	W
	Mancy, K. H., Editor, *Instrumental Analysis for Water Pollution Control*, Ann Arbor Science Publishers, Ann Arbor, Michigan, USA, 1971 (pp. 331). (Based on a manual for a training course at Warsaw, Poland, 9-21 June 1969, organized by the Regional Office as part of the UNDP/WHO		E	W

TABLE 1 (cont.)

Ref. no.[2]	Title[3]	Place and date	Language[4]	Subject[5]
	assisted project POLAND 0026)			
	Krenkel, P. A., Editor, *Automatic Water Quality Monitoring in Europe,* Department of Environmental and Water Resources Engineering, Technical Report No. 28, Vanderbilt University, Nashville, Tennessee, USA, 1971 (pp. 487). (Proceedings of a seminar at Cracow, Poland, 29 March - 2 April 1971, organized by the Regional Office as part of the UNDP/WHO assisted project POLAND 0026. See also report EURO 3119)		E	W
EURO 3109(1)	*The Hazards to Health of Persistent Substances in Water: Technical Documents on Arsenic, Cadmium, Lead, Manganese and Mercury,* WHO Regional Office for Europe, Copenhagen, 1973 (pp. 159). (Annexes to report EURO 3109(1) on a working group meeting at Helsinki, Finland, 10-14 April 1972)		E F R	E
EURO 3402(2)	Patrick, P., *Model Code of Practice for the Disposal of Solid Waste on Land,* WHO Regional Office for Europe, Copenhagen, 1973 (pp. 32). (A chapter of the Manual on Solid Waste Management under preparation by the Regional Office. See also report EURO 3402(2))		E F R	S
	Deininger, R. A., Editor, *Design of Environmental Information Systems,* Department of Environmental and Industrial Health, School of Public Health, University of Michigan, Ann Arbor, Michigan, USA, 1973 (pp. 98). (Proceedings of the workshops and country statements of a seminar at Katowice, Poland, 11-20 January 1973, organized by the Regional Office as part of the UNDP/WHO assisted project POLAND 3102)		E	I
	Ditto, Ann Arbor Publishers, Ann Arbor, Michigan, USA, 1974 (pp. 429). (Proceedings of the individual papers)		E	I
EURO 3404(1)	Fish, R. A., *Toxic and other Hazardous Waste,* WHO Regional Office for Europe, Copenhagen, 1974 (pp. 30). (A chapter of		E	S

TABLE 1 (cont.)

Ref. no.[2]	Title[3]	Place and date	Language[4]	Subject[5]
	the Manual of Solid Waste Management under preparation by the Regional Office. See also report EURO 3404(1)0			
EURO 3110(2)	Ekedahl, G., Rondell, B. and Wilson, A. L., *Analytical Errors*, WHO Regional Office for Copenhagen, 1975 (pp. 56). (A chapter of the Manual on Analysis for Water Pollution Control under preparation by the Regional Office. See also report EURO 3110(2))		E	W

[1]A complete list of reports which covers the entire activities of the Regional Office for Europe for 1949-1973 is available on request.

[2]The reports are available free of charge from WHO Regional Office for Europe, 8 Scherfigsvej, 2100 Copenhagen, Denmark. When requesting please quote Reference Number only.

[3]The reports are a result of one of the following activities: (A) = contracted from author(s); (C) = Conference; (EG) = Evaluation Group; (WG) = Working Group; (PM) = Planning Meeting; (Sp) = Symposium; (Sm) = Seminar; (SC) = Steering Committee.

[4]The reports are available in one or more of the languages indicated: (E) = English; (F) = French; (R) = Russian.

[5]Subject indicator column for fast reference: (A) = Air; (E) = Ecology; (F) = Food; (I) = Information; (M) = Manpower and Management; (N) = Noise; (P) = Pesticide; (S) = Solid Waste; (R) = Radiation; (T) = Training; (W) = Water.

[6]Publications not published by this Office can be requested direct from their publishers or bookstores.

WHAT MUST BE UNDERSTOOD BY CRITERIA, NORMS AND STANDARDS

for A.R.M. Lafontaine

by Dale R. Lindsay

The United Nations Conference on the Human Environment held in Stockholm in 1972 has decided that the establishment of health criteria and of resultant primary protection norms is one of the essential elements for the design and implementation of practical programmes in environmental health and particularly for the control of the quality of air, water, food and the working environment.

That decision has been confirmed by the World Health Assembly Resolution 24-27 endorsing the Director General's proposal for a long term programme in environmental health stressing the need to establish and promote international agreement on criteria, guides and codes of practice with respect to known environmental influences on health, with particular emphasis on occupational health exposure, and water, food and air wastes, and to obtain further information on levels and trends of these "and the need" to extend the knowledge of effects of environmental factors on human health by collection and dissemination of information, stimulation, support and coordination of research and assistance in the training of staff.

The purpose of such resolutions was the protection of man against identifiable, immediate and long term effects on his health and well-being arising from the exposure (singly or in combination) to various environmental factors. An additional objective is the provision of guidance on levels or conditions of exposure that would be consistent with the protection of the health of exposed population (taking into account the most sensitive members of the public).

Those objectives may be achieved only if the following steps have already been realized:

1) to settle the nature and the origins of the pollutions and harmfulnesses;
2) to assess the magnitude of the hazards for man's health and mankind, related to direct effects on health or to effects of man secondary to alterations of the human environment;
3) to fix the short, middle and long-term goals;
4) to establish a monitoring system in order to measure the exposure levels and their related risks and to institute an epidemiological surveillance with the aim of a better knowledge of the hazards and of the efficacy of the

preventive or corrective actions.

But before fixing any methodology, it is important to have clearly in mind the definitions of the terms in use, and to avoid the confusion which has been surreptitiously introduced in the terms leading to disturbances in the actions at national and international levels and misunderstanding between scientists and administrations.

The purpose of this short paper is to reach a better agreement in the usage and the interpretation of the terms and in the standard-setting process.

It may be useful, in the first place, to repeat the meaning of the term "Health" for WHO: a state of complete physical, mental and social well-being and not only the absence of disease or disability. The term "well-being" includes the notion of a dynamic harmony, which can differ in space and in time, giving to the man the feeling of being satisfied with his condition and his environment: it includes a quantitative approach emphasized by L. Breslow (Inst. J. of Epidemiology, 1972, -1-347-354).

In other respects, the environment must be defined as the entity of the physical, chemical, biological and sociological circumstances, external to them, that the live beings and especially the man, encounter and that intervene in their present condition and in their evolution either to promote them or to slacken or hinder them. Any deterioration of the normal environment may influence negatively and generally impair the health, but defined environmental condition may also be favorable or at least no harmful to man's health and well-being. At the present, only harmful effects are too often considered and I think it is of fundamental importance to consider simultaneously

1) the relationship between exposure and different effects, not only the harmful ones, but also the harmless or propitious consequences;
2) the concept of exposure limits that will ensure, in an unconditional or in a conditional way, the protection of human and environmental health.

The first step to reach these concepts is to evaluate the available scientific data which will make possible the establishment of a link between exposure to be identified, specific or combined, environmental factors and their effects. The link should be expressed, if possible, in terms of numerical signs of exposures to environmental conditions (physical, chemical, biological, social and cultural) on the one hand, and the consequences, deleterious or not, of such exposure to man's and mankind's health (expressed also numerically as far as possible on the other hand: generally the receptor or target will be clearly defined).

The relation, ideally expressed numerically, represents a criteria which is a way of judgment (Krinein, in Greek, means to judge), designating a ratio scientifically demonstrated, between exposure to an agent, a pollutant or another factor of agression and the risk or magnitude of effects, generally undesirable, underspecified circumstances defined by environmental conditions and targets. Ideally, criteria or complete sets of quantitative exposure/response relationships for all environmental factors for different effects, and for different population groups, covering whole ranges of exposure levels. But there are innumerable difficulties in establishing a cause-effect relationship and still more a quantitative relationship, particularly for chronic or delayed effects or for combined or cumulative exposures.

However, although the available number of criteria is very limited, the existing ones are very useful: but when they are used, the following rules must be respected:

a) a criteria is always a scientific judgment and never an exposure limit; it is excessive to consider a criteria as an exposure limit or to speak of criteria when we consider a no-effect level (a no-effect level must, as a matter of fact, be considered as primary protection standard);

b) a criteria is a direct relationship between an exposure and an effect on the target; we must be careful to use the term criteria when speaking of indication of pollution.

The relationship between the atmosphere concentration in sulfur dioxide itself and health effect is difficult to be accepted as a criteria because on one hand the deleterious effects are in connection with the combined pollution "SO_2 + particulated matters" and not only with the level of sulfur dioxide and on the other hand, the reduction of the level of SO_2 does not necessarily demonstrate reduction of the global.

The objectives for quality are referred to the environmental conditions: they designate the bundle of requirements to which the environment or a well defined part of it must satisfy at a given moment, in the present or in the future.

In establishing such objectives, will be considered:

a) an exposure limit for a basic security (the human being or the target may not be exposed to an unacceptable danger);

b) a no-effect level limit: under that limit, no effect may be detected. This concept remains theoretically doubtful because it is frequently impossible to distinguish between the exposure-response curve which shows a true threshhold and one where the response follows an asymptotic relationship.

c) the criteria must take into account the characteristics of the exposure (intensity, frequency, variability, rate, duration) and of the biological response (prompt response, subacute, chronic) and the variety of the host factors: the delayed effects (including possibility of mutagenesis, carcinogenesis and teratogenesis). At the same time, the host factors will be considered and especially the vulnerability of special groups (children, child-bearing women, old people, persons affected by disease: this vulnerability may be permanent or temporary, inherited or acquired).

The criteria are normally the first step to set objectives for quality (and exposure limits) on a scientific basis: however, the criteria may not exist but this absence is not an hindrance to establish on partial knowledge practical limits and quality requirements useful to build actions and regulation.

Biological experiments or epidemiological researches can never be sufficiently precise to prove or to disprove with certainty of a threshhold: however, in practice, the concept of threshhold and of no-effect level is nevertheless useful even if it represents a crude and artificial simplification in most cases.

The objectives for quality and the exposure limits are, when possible, based on the admitted criteria which, therefore must be established and accepted at the highest levels of the international organization. In the choice of such objectives other factors outside the scientific judgment, may be used as a basis of the criteria: i.e. special regional conditons or balance between advantages and disadvantages for man and mankind. They may also be established in absence of criteria: they will then take into account all the available knowledge including experimental work, and all the valuable comparisons with comparable agents or substances.

We insist that the objectives for quality are goals and not ways of reaching these goals: they are the purposes towards which we guide all our efforts, and not the practical, technical or legislative measures to control the situations.

The objectives for quality and the exposure limits (also called primary protection standards) will lead to secondary standards or norms (including the working limits): again, such secondary standards or norms may also be the result of more arbitrary decisions where neither the scientific knowledge nor the experience have been able to determine criteria or to choose objectives for quality: these norms are established with the objective of limiting or preventing the exposition of the target, the man, or other components of the environment: they are means through which the objectives of quality may be reached or approached.

These norms apply directly or indirectly to members of the public or to official or non-official responsibles. Generally, they determine levels of pollution or annoyance, levels which may not be overstepped in a section of the environment, in a target, in a product, etc. The norms may be established, on one hand, by the responsible authorities, by means of laws, regulations or administrative prescriptions or, and on another hand, by mutual agreement, or by spontaneous acceptance: very often the norms may appear under the form of a code of practice which contains all the usual customs and spontaneously accepted rules: such codes offer generally a realistic solution to a lot of problems.

The norms, which are often called standards, may be directly bound to the components of the environment, but they may also be applied to the products (in the largest meaning of the term and including packaging, labelling, instructions for use, etc.) and to the operating processes (limits for industrial atmospheric or liquid emissions, regulations for conception or construction of the plants, specifications to be observed during production).

Conclusions

We have repeated well known things, but we have thought that it was not unnecessary to describe again the different phases of the intellectual process leading to standard setting at either national or international levels. Ideally, the first step is the choice of the criteria, where they are available and evaluable: in many cases, we are before vacuum or deep uncertainties and more research at the toxicological and epidemiological levels is urgent. At the same time, an international effort is required for the choice of a strict previously established programme of control of the new product or of the new processes: problems such as these of the PCB's or of the vinylchloride should be avoided in the future.

From the criteria, it is possible to deduct objectives of quality for human health (and well-being and for man's environment). These objectives must be based on all other available informations when criteria do not exist: they are always temporary and must be adapted to the progresses in knowledge and to the changes in the other variables and interactions which must be considered in the establishment of an advantages-disadvantages balance (erroneously too often called a risk-benefit balance). We must remember that the overall aim of the objectives of quality is to give population groups the benefits of a technological society while at the same time minimizing the detrimental effects of environmental pollution and the dissemination of the hazards.

To respect the objectives of quality, different practical actions going as far as interdiction have to be taken. They may be very different, varying in the geographical, demographical, cultural and economical conditions and varying with the realistic feasibilities and the amplitude of the hazards. Many of these practical actions will be based on norms, on derived working standards. In certain circumstances, such actions may be based on provisional information if no criteria exist or no objectives have been defined: the suspicion of eventual hazards of man's health or mankind is a sufficient ground to justify protective measures. Such a standard-setting process in environmental health implies a permanent collaborative effort between the international organizations, the national health authorities, the research units and the involved plants or administrations to identify the risks and the hazards and their origin (including new pollutants and new uses of pollutants), to appraise and evaluate the health and environmental effects reviewing the experience obtained elsewhere, to consider the social and economic implications and the general policy and to refine the objectives and the norms as the knowledge increases.

I apologize if I have introduced more general problems in a forum especially devoted to water quality, but the criteria, objectives of quality and norms to be applied to water can illustrate the process I have tried to explain.

Finally, two remarks:

The first one, already raised when we have evoked the criteria, refers to the use of indices. We have given the example of sulfur dioxide as index of air pollution, we can add that of the colicounting to appreciate the quality of the waters. Such indices may be very valuable for routine surveillance but otherwise are of no more significance than the recording of body temperature for the diagnosis of a disease. Exceeded indices nearly always mean that circumstances are unusual and possibly (not inevitably) merit concern. But the fact that the indices are not exceeded must not exclude other methods of surveillance based on working limits specific for a certain factor (as salmonella in water) or more sophisticated or more intensive surveillance.

The second remark referstto the "working limits" or "secondary standards" used for practical application of the norms, i.e. the exposure limits set with respect to specified media other than the receptor as water, air, soil or food, and designed to ensure that under specified circumstances, the norm, if available, is not exceeded (tolerances of contaminants in food, permissible levels of airborne toxic substances in occupational exposure, etc.).

These working limits have often been set before availability of the criteria and before the establishment of objectives of quality: sometimes they are fixed before any effect on health or on environment have been observed, taking into account the experimental data.

THIRD DAY

OPENING REMARKS

DR. COULSTON: Mr. Chairman, Ladies and gentlemen, because of the wonderful hospitality shown to us by the National Wildlife Federation, particularly through the good offices of Dr. Kimball, we decided to create a resolution which this body should approve, to write to Dr. Kimball and thank him in the name of the Forum, for the services and contributions they have made towards the meeting. Therefore, with your approval, by a show of hands, this shall be a unanimous decision, and such a letter shall be written.

We wish also to thank Mr. Tom Riley and his staff. And especially the ladies outside, Ms. Nancy Quigley and Ms. Rose Kobryn, for their splendid assistance in making the Forum a success.

DR. MRAK: As Chairman of this morning's session, I would first like to ask Professor Truhaut to introduce a very distinguished gentleman from France.

DR. TRUHAUT: It is my pleasant privilege, Ladies and Gentlemen, dear colleagues, to introduce Professor Marois, the distinguished President of the Institute of Life in France. He is an eminent biologist, especially interested in endocrinology, histology, cytochemistry. Certainly, all of you know him because you know scientists all over the world. But you can imagine how happy I am to welcome a countryman and friend like Professor Marois. A warm welcome to you, Professor Marois.

DR. MRAK: Thank you very much for coming to the Forum, Professor Marois, and I hope you will feel free to participate.

The first speaker this morning is Dr. Smeets, who will talk on "The Conceptual Basis of European Drinking Water Standards."

THE CONCEPTUAL BASIS OF THE EUROPEAN COMMUNITIES' DRINKING WATER STANDARDS

J.P.M. Smeets

I. Introduction

On 22nd July, 1975, the Commission submitted for approval to the Council of Ministers of the European Communities a draft directive relating to the quality of water for human consumption. The objective of this regulation is the protection of public health by establishing standards to be respected by the nine member states. They concern the quality of drinking water supplied by public mains systems, water stored or delivered in bottles or other containers (but not mineral and medicinal waters) and water used for the production of foodstuffs and drinks.

The submission of this proposition was requested by the Council of Ministers in the Action Programme of the European Communities on the environment, approved by them on 22nd November, 1973. The chapter dealing with the measures to reduce pollution and nuisances provides for the setting of standards for toxic chemical substances and for germs which endanger health and which are present in water intended for human consumption. It provides also for the definition of physical, chemical and biological parameters to be considered and corresponding to the different uses of such water and in particular to drinking water.

II. Objectives

Because of the increase in water requirements, it is necessary to draw on all potential sources of water which can be processed for human use. In particular, surface waters are increasingly used for this purpose. Since they often contain non-degradable polluting substances, these waters must be subjected to increasingly elaborated purification processes. The quality of water supplied for human consumption must be supervised, and levels of toxicity and noxiousness fixed with reference to the most up-to-date scientific knowledge in this field.

A draft directive on the quality of surface waters intended for the production of drinking water was already approved by the Council in November 1974.

Inside the European Communities, the national legislations relating to the quality of drinking water differ from one Member State to another. In fact, one finds a certain number of provisions applicable to drinking water, which are neither comprehensive nor at the same stage of planning and development. The recommendations of the World Health Organization have only been used in a few cases as a basis for national legislations. These discrepancies can be an obstacle to trade within the Community and can therefore have a direct bearing on the functioning of the Common Market.

Since the WHO non-mandatory standards were proposed, and despite a recent revision, the significance to health rightly attached to the presence of metallic ions in drinking water has increased considerably. It is now much more important than was envisaged when the standards were drafted.

The same remark could be made with reference to other groups of substances, in particular organic and organo-metallic micro-pollutants.

Furthermore, another very important factor is that the properties of the water available are often altered by the consumer with a view to protecting his domestic supply system. This alteration was not considered in the recommendations made by the WHO.

There is a remarkable growth in the use of softening systems to modify the composition of water supplied to the consumer. This is due to the pressure of changes made in the methods of satisfying normal demand and it is taking place in a partial legal vacuum because of the inadequacy of existing regulations. In view of the equipment and products being used, this question is at present as important, on an international scale, as that of the treatment of groundwater in containers made of materials of varying stability.

Considerable reserves of groundwater are also being drawn on, supplied on a commercial basis and exported to countries of the European Communities and also to non-member countries. This is particularly true of table-waters. In recent years, these exchanges have raised questions of a technical, legal and medical nature. So long as the countries remained autonomous as regards water supplies standardization, often only meant the solution to a domestic problem. Today the situation is changed; because of the increase in demand, associated with population growth and new habits and requirements, former sources like groundwater are insufficient; surface water must be used. Such water, however, has many uses and the rivers and streams must henceforth cope simultaneously with widely differing requirements. Where international rivers are concerned, it evidently becomes necessary to compare the measures needed and to coordinate facilities for appraising the situation.

To sum up, the enactment of this mandatory directive is in keeping with general rethinking on the subject of water quality. It is intended to reconcile for the nine Member States as a whole and in a standardized way the conflicing needs of productivity on the one hand and public health on the other. These needs are linked with the necessity to use surface waters which must serve several purposes simultaneously, in particular navigation and the disposal of industrial and urban effluents.

III. Technical aspects

For the establishment of the draft directive the recommendations made by the WHO were considered by groups of experts of the Member States as well as the recent studies on the toxicity of heavy metals and the effects on health caused by microbiological factors. Analytical measurement techniques were studied and compared between laboratories. A scientific European colloquium was organized on the relations between "Hardness of Drinking Water and Public Health," (11-13 May, 1975, Luxembourg).

The 62 parameters selected for standards form a coherent whole on the basis of which the properties of water intended for human consumption can be logically defined. Toxic substances and noxious germs are given priority.

The parameters are classified in five classes:

1. organoleptic factors (colour, turbidity, odour, palatability, temperature)
2. physicochemical factors (pH, conductivity, total mineral content, total hardness, calcium, magnesium, sodium, potassium, etc.)
3. biological factors (dissolved oxygen, oxidability, BOD_5, total carbon)
4. undesirable or toxic factors (about 25 metallic, non-metallic ions and organic compounds)
5. microbiological factors (Coliforms, Streptococci, Clostridium, Salmonella, pathogenic Staphylococci).

The radiological parameters have been dealt with in the framework of the Euratom-treaty basic safety standards for protection against ionizing radiation.

The choice of parameters was based on several criteria; they relate to:

- essential health requirements; in this connection, Maximum Admissible Concentrations (MAC)* were fixed for all

* MAC: the concentration below which a substance in water cannot, in the course of continuous ingestion, cause or directly or indirectly result in, an identifiable effect harmful to health in a statistically representative sample of the population involved.

pollutants and Minimum Required Concentrations (MRC)* were laid down for calcium, magnesium, bicarbonates, chlorides and sulphates.

- the need to consider special local situations (climate, hydrogeology), and the concern of the responsible authorities to be able to take appropriate action in exceptional circumstances (natural catastrophe, floods). To this end it is possible to incorporate Exceptional Maximum Admissible Concentrations (EMAC)**.

- the wish to improve the quality of water intended for human consumption. The Guide Levels (GL)*** chosen represent target quality objectives.

It is necessary to define and agree on terminology to be used within the Community in order to eliminate the ambiguities of terminologies at present in use both at national and international levels.

The considerations underlying the choice of the different parameters are based on present scientific knowledge of the effects produced by water pollutants or by substances contained in water on the population in general or on specific population groups (children, aged persons, the sick, etc.).

Scientific knowledge relates both to the immediate effects and also to the long-term consequences. Since there are still considerable gaps in our knowledge in this area, e.g. with respect to the cause-effect relationships, it is necessary to be very cautious in deciding what levels to select.

The updating of technical and scientific knowledge will necessitate a five-yearly revision of these standards. Moreover, a request for a partial revision may be made, either by a Member State or in answer to a proposal from the Commission, particularly with reference to Exceptional Maximum Admissible Concentrations; these must be of a <u>temporary</u> nature.

With reference to the <u>monitoring</u> of these standards,

---

* <u>MRC</u>: the minimum concentration of a substance, the presence of which is essential for preventing the occurrence of identifiable harmful effects in a statistically representative sample of the population involved, either directly or indirectly, as a result of repeated ingestion.

** <u>EMAC</u>: the exceptional maximum admissible concentration which may be authorized locally by the relevant authorities, either temporarily, in view of particular meteorological conditions, or permanently, in view of geographical or geological conditions.

*** <u>GL</u>: the concentration of a given substance in water which it is advisable not to exceed.

representative sampling and a recognized system of analysis should ensure that meaningful and comparable results are obtained.

Spot sampling, usually only single samples, on the basis of which multiple tests are carried out, should be abandoned in favour of multiple sampling on which only a few tests, and not all, are based.

The size of the population involved and the capacity of the source of supply should also be considered.

Moreover, the consumer, in order to ensure his own amenity and to protect his domestic distribution system, sometimes modifies his water supply. Since this modification has an effect on both health and the safety of installations, it is essential that the sampling after rinsing which is normally practiced, should be preceded by an initial sampling of water which has stagnated in the pipework.

There is, therefore, an order of priority among the parameters determining the frequency of sampling and analysis. This order of priority is taken into account in the system of standard analyses already in use in certain Community countries. Three types of analysis, of increasing complexity, are selected, varying from

(A) constant monitoring of distribution networks supplied from underground water or surface or mixed water, subdivided according to the size of the supply; to

(B) systematic periodic monitoring, whatever the origin of the water might be to supplement the constant monitoring; and

(C) occasional monitoring in special situations or in case of accidental circumstances.

The frequency of these standard analyses is determined by two essential factors:

1. For analyses A and B, the capacity of the installation and the size of the population served,
2. For analysis C, the source vulnerability and the hazards threatening it, regardless of the size of the population served and the capacity of the installation.

With reference to the techniques themselves, a number of reference methods have been suggested which take account of the most recent technical advances and of conditions found in some laboratories which are not yet adequately equipped.

In this respect, we have performed a number of interlaboratory comparison programmes for the quality control of analytical techniques and to arrive at reference methods. Furthermore, we had a workshop during one week in June of this year with thirty very qualified experts of our nine Member

States to improve and harmonize the microbiological measurements.

IV. Some additional comments

The fact that standards have been proposed does not mean that our knowledge, in regard to all the parameters considered, is complete from either the toxicological or the technical viewpoint; many problems have still to be solved. When newer data becomes available, they will be considered every five years.

It might be of interest to summarize the main problems which are being considered now. For some of them, the Commission has already concluded study contracts, others may follow.

Since drinking water is monitored in general immediately after preparation in the treatment installations, one should have more information on the modifications which take place in the produced water

(A) during the shorter or longer period in which it lies overnight in the pipelines, or

(B) depending on the distribution system.

Modification of the concentrations of e.g. Fe, Zn, Pb, Cu, Cd, hydrocarbons and dissolved oxygen, with reference to the pipeline material used, should be considered.

Only rough assessments are available concerning the daily drinking habits of the consumer (quantities used for drinking water and food preparation, frequencies, use of bottled water, etc.).

There appeared to be a great lack of information concerning the qualitative and quantitative relationships from the physiological and metabolic point of view between the uptake of minerals and other essential elements with food and drinking water. In this respect, we noted a great disagreement between nutritionists.

It has appeared that the surveillance and monitoring of the different parameters possibly present in drinking water is in many cases far from being systematic; this is particularly true for the microbiological analyses. Such a monitoring is very costly but, on the other hand, there is often a lack of sufficient control.

With respect to the relationship between the hardness of drinking water and public health and the proposals put forward for minimum required concentrations for some parameters, it might be of interest to summarize the conclusions of the scientific colloquium we have had, from 11-13 May in Luxembourg.

There seems little doubt that there is some factor or factors associated directly or indirectly with the supply of drinking water which is of vital concern to public health. Whether we are dealing with a protective or injurious factor, whether it is a higher or a lower concentration which makes it injurious or protective, whether it is a single factor, a series of factors, or a combination of factors interacting with each other, we do not yet know. Some epidemiological surveys performed on a regional basis have shown an inverse statistical association between hardness of drinking water and mortality, in particular cardio-vascular mortality. The complexity of the situation and the almost total ignorance of the human requirements of the various elements and the equally great ignorance regarding the functions and possible toxicities of the trace elements, compound the difficulties. For the wide range of trace elements there is no reliable knowledge of their levels in water or food, and no information on their degree of absorption.

It was established that some ten elements need to be investigated intensively. These, in alphabetical order, are: calcium, cadmium, chromium, copper, lead, magnesium, selenium, silicon, sodium and zinc.

In the current state of knowledge, artificially softening water cannot be encouraged. It is also possible that the mixing of hard and soft waters could have future consequences for public health. Where alterations in hardness of drinking water are anticipated in towns with a population in excess of, say, 50,000, studies should be performed before and after the alteration and control towns should be included. A priori, those independant and dependant parameters to be considered in such studies should be specified and coordinated under a central body.

It is evident that there is a need

- to have a minimum required concentration of calcium and magnesium,
- to take into account the important roles of both sodium and potassium,
- for studies on the relative importants for the transport of elements of water in relation to food,
- for studies of the individual in relation to changes in drinking water of differing qualities,
- for post mortem studies to provide information regarding the absorption of various trace elements during long periods of exposure to water of differing qualities.

V. Conclusions

The submission by the Commission to the Council of Ministers of a draft directive with standards establishing

- the choice of a series of parameters
- the numerical values given to them

- and the measures relating to the monitoring and supervision, and so establishing the quality of water for human consumption, is an important step at Community level to protect human health. The conceptual basis of these proposals which envisage a harmonization of legislation in this respect between the nine Member States is based on the most up-to-date scientific knowledge in the different fields which are concerned. The task was difficult, since there are still many unknown factors, toxicological and physiological, as well as technological.

We are very grateful to all the experts who have collaborated with us intensively and in an harmonious way over a period of two years to prepare this proposal. We all realize that still more and better information and scientific data are needed to perfect the work performed. In this respect, we shall go on with our studies, so that the five-year revisions foreseen will result in another important step.

LATIN AMERICAN WATER SUPPLY

David Donaldson

In keeping with the theme of this International Water Quality Forum, the Pan American Health Organization has been asked to describe the Latin American water situation. To do this, we propose to use the experiences of the region of the Americas in order to help you understand how water quality is handled in a developing country situation.

First, a few facts and figures so that you understand some of the characteristics of the countries that I will speak of. The area that I am going to talk about is the Americas, outside of the U.S.A and Canada. The area I refer to covers 27 countries. There is a population of about 300 million people. It is divided in urban and rural characteristics of about 125 million in the rural population and about 178 million in the urban population. Our largest country is Brazil, a country which has 103 million people. The per capita income of this area ranges from Argentina with $800 per year, Brazil at about $250 per year, Haiti which has $70 per capita per year. For comparison, Canada has a per capita income of around $2,380.

Over the last fourteen years, we have supplied water to about 103 million people, and probably have had the most progress in the water supply area outside any of the industrial countries. This is as if we had built enough water supplies to supply the total population of Canada five times over. We have currently supplied 57% of the people of the region. 80% of these are supplied from house connections.

We have spent about $4.9 billion. Of this, about $4.2 billion has been spent in the urban and the rest in the rural areas. The money has come from a number of sources. It has mainly come from the countries themselves. The Inter-American Development Bank, the World Bank and the Canadian Development Agency, have all supplied seed money for this program.

The strategy used in developing this program was set down in the early 1960's by the heads of the governments when they met in Ponte del Este. They established some coverage goals. They said that during the 1960's they would provide water supply and sewage to 70% of our urban population, and 50% of our rural population. The strategy that came out of those goals was that we attacked first the large cities. We were concerned first with quantity.

The progress that has been achieved, based on official country estimates, show that by the end of 1974 about 172 million people in the urban areas enjoyed what we call a potable water supply. 79% of these enjoyed it from house connections and only about 34 million of the rural inhabitants (some 27% of the area of the people in the rural area) enjoyed a safe water supply.

In 1972, the Ministers of Health met and developed a new set of goals for which we are currently developing a new strategy. The new goals set down, again, urban and rural water coverages. For example, the urbanite says to provide water supply through house connections to 80% of the population, or, as a minimum, reduce those without water supply at the beginning of the decade by 50%. We have a similar goal, which is 50% for the rural population.

The problems that were faced in trying to establish these goals were that this region has one of the highest growth rates in the world. Also, we have explosive growth in our major cities. We have almost a six percent rate of growth in some of the cities. In other words, we must duplicate the water supply approximately every twelve years.

We have taken this strategy, we have taken these goals, and we have studied the population patterns. We can now say that between the time period of 1971 to 1981, we know that we have to supply 156 million people. We have to spend approximately $4.4 billion. In the face of that, our strategy has changed. When I say our strategy, I should clarify that it is the country's strategy. We often times speak of ourselves as the countries. The strategy of the countries have now changed to a four-pronged attack. They have to continue improving coverage by building new systems. They have to upgrade existing systems, for example, 34 million people in the urban areas only have an easy access to water. We now have to start improving the water quality in existing and new systems. We have to develop the institutions, the criteria and the standards that will allow the financial operation and management of these services on a long-term basis.

The approach that is being used is that we must respond to a variety of population groupings. We have dispersed population, we have villages, we have towns and cities, slums in major cities. And we have to respond to a large number of human problems, basically gastrointestinal diseases and lack of acceptable sanitary conditions. The response that has been developed to this are basically four coordinated programs.

The first program is a program that responds to this condition, the dispersed population. The next program, is one which begins to cover the villages and the towns. The next one covers problems such as a population of 7 million people in the metropolitan area of Sao Paulo. And finally, the last program tries to respond to the fringe areas of the various countries.

Our general approach is to first promote the provision of

water and sanitary facilities to those living in the village areas and in the outskirts of the major cities in such quantities and in such a manner that it will allow the fulfillment of at least survival public health functions. Once these conditions have been fulfilled, the services should be extended in the metropolitan areas to those who only enjoy easy access and to the less concentrated populations.

We are always striving to improve drinking water quality. That gives you an idea of the scope of our multi-faceted problem. Let's take a look in more detail at what is happening in the country - what are their needs and what are they trying to do to solve them.

One of the major countries of the region has mounted a survey of water quality of all of the country outside the capital city. This was a survey of approximately 300 communities, and covered towns from several hundred thousand people down to about 500 people.

Six percent of the population served by the aqueducts of this country receive contaminated water on a consistent basis. The problem is fundamentally the lack of operation and maintenance, and in good part, a lack of sanitary education of the people involved in operating these systems. A good number of the aqueducts, though not in the majority, required the installation of equipment for the minimum treatment of water, that would mean chlorination. For the aqueducts that reported contamination, corrective measures were generally inexpensive and easy. It is interesting to note that in this country, no mention was made of anything but bacteriological quality.

In an examination of the report, it is interesting to note that there appeared to be at least three water quality standards that are unofficially applied in a country which has one officially adopted water quality standard. These three are a set of water qualities for the dispersed area. You have no gross problem; you do not have people dropping dead when they drink the water. There is another series of standards that are applied to the villages and to the fringe areas which cover basically bacteriological quality and some of the more gross chemical qualities. You usually have a bacteriological chemical analysis made of the water supply when it is installed, and then at some future dates, you have some assessments made of the quality found at that time. Then, you have what is the official water standard of the country, and this is generally only applied in the major cities.

Now, one asks, "WHY?" Dont't the official believe in the need for high quality water? Don't they understand the problem? When the developing countries of the world were asked what were the reasons that they built water supplies, the answers are often quite surprising. A recent WHO document,

"Community Water Supply and Excretives Disposal Situation in Developing Countries," has the following statement, and I quote, "It is interesting to note the criteria pattern in the region of the greatest achievement," (they are talking of the region of the Americas) "shows health as the next to lowest priority in terms of frequency of mention and willingness of community to participate as the next to highest priority, while safe community water has a positive impact on health, could well be undertaken for other reasons." In interpreting this statement, you have to realize that WHO's definition of health is not the near absence of disease, but the total physical and mental well being of the human being. While we have the above mentioned statement, one cannot ignore the facts and figures regarding mortality and morbidity. During 1969, 1972, according to published figures, El Salvador, a country of approximately 4 million people, had 197,000 cases of bacillary dysentery and 11,750 deaths. These are registered cases and are probably on the low side.

Yesterday, I pointed out some of the other facts and figures, but again, to quote from one of the official PAHO (Pan American Health Organization) documents in this case, "severity of health problems arising from a lack of potable water supplies and sewage systems, stand out clearly in an analysis of mortality and morbidity data from diarrhoeal diseases in countries of Latin America, especially among children under five years of age. It shows that the group of enteritis and other diarrhoeal diseases in 1972 was one of the five principle causes of death in 19 of 34 countries. In five countries, this group was the leading cause of death, and in another six, the second leading cause of death.

Among children under one year of age, diarrhea appeared as one of the five principal causes of death in 32 countries. In other areas, it was first place in ten countries, second in nine. Moreover, in the age group one to four years, diseases ranked among the first five causes in 27 of 32 areas, first in fourteen and second in six. For those under one year of age, the highest rate is 2,500 per hundred thousand. This is compared against 27 for the U.S.A. In other words, you have a rate approximately 100 times what you have here in the U.S.A. In the one to four age group, you have rates which are approximately 700 times what they are here in the U.S.A.

With that as background, we have asked ourselves, what has our experience taught us? We feel that our solutions to the water quality problem cover an enormous range of situations. We go from the problems of simple chlorination of a single source, to the most advanced problems, long low-level effects of mercury, such as the problems in Sao Paulo. All of these take place in one country which has very limited resources.

Accordingly, our strategy must be a very flexible one, because we are dealing with a highly dynamic problem. As we solve the problem, the problem changes. We must be very cost-effect oriented in our solutions. To do this, we are making use of multidisciplinary (by this, I mean technical, financial and medical teams) to examine the problem and evaluate resources for each level.

The general approach we are using can be best visualized as a series of a 15 cell matrix. Down one side, we list the various areas, including the dispersed, the villages, the medium sized, the fringe areas and the metropolitan areas. Across the top, we are listing the degree of difficulty in the resolution of the medical problems and the financial and the technical problems. We have to develop one of these for each condition that we are trying to confront. It's solution will vary as the response is made to the problems. In other words, as we bring resources to bear on the problem, our matrix changes. Therefore, the standards that can be developed at any time would depend on the integration of the various matrix factors, one per disease, per health problem. This is the only way we can come up with results that are realistic and enforceable.

It is well to note that in this manner, a water quality standard becomes a flexible response to an environmental problem. As one changes from biological causative agents, the classic water borne diseases, to the chemical causative agents, we feel the standards will have a longer life. For example, in Bolivia, the major problem is enteric diseases. Their response is to develop the resources which may be staged to reduce the level to an acceptable one.

In Washington, D.C., which is at the far end of this dynamic process, the problem is to maintain enteric diseases at a low level. This approach is being explored in greater depth at a conference that PAHO will be holding on the control of drinking water quality later this year in Sao Paulo. At this time, decisional level officials from the monitoring agencies and ministeries of health, will look at the medical aspects, production agencies will look at the technical aspects, and the national planning boards will consider the financial developmental aspects. We will meet to develop guidelines for plans of action that will respond to the various levels of standards that are needed to maintain the development of the countries needed that they are realistically able to maintain.

In closing, we must remember that the basic reason for a water quality standard is to protect human beings. If he is dying of diarrheal disease, there is no sense in worrying about a long-term level effect of mercury. But our experience shows the standards must take into account the fact that we are dealing with a highly dynamic program, with a highly dynamic system. We must be prepared to change as the situation changes. Our experience has shown, in developing countries, that for practical reasons one often has to develop a series of multi-level sets of standards that respond to the different resources found in the different areas of the country.

While major environmental problems in Latin America are still biological, the chemical problems are increasing in number and in location. To respond to this, the countries are developing these multi-level standards, using multi-disciplinary teams of experts that include the financial, the technical, and the medical expertise.

In view of the extremely wide differences between the countries and their needs, a wide range of solutions is being sought. We have found water quality in Latin America to be a highly complex and challenging field. We look forward to your help, your assistance, and your guidance, in seeking to reduce the time that is required to bring safe water to all of the inhabitants of this region.

WATER PURITY PROBLEMS IN JAPAN

Iwao Kawashiro

In our country, the quality control of drinking water, including natural fresh water from well and municipal drinking water, falls under the jurisdiction of the Ministry of Health and Welfare. And our Institute, National Institute of Hygienic Sciences, which belongs to the Ministry, is responsible for the technical and scientific field of the drinking water quality control.

It is a matter of fact that we also have the official drinking water quality standard which was laid down in consideration of the recommendations from WHO and the American standard as well as other countries.

On this occasion, I would like to mention some of the problems regarding the Japanese water quality standard which we are now confronting.

The first problem: regarding cadmium level in drinking water, we have a tentative level of less than 0.01 ppm, as is seen in the U.S.A. and WHO standards.

But, when the WHO Working Group on the Hazards to Health of Persistent Substances in Water held its meeting in Helsinki in April 10-14, 1972, the Working Group had recommended that judging from 70 μg recommended by WHO and FAO as the ADI of cadmium for adults, the limit of cadmium in drinking water (0.01 ppm) should be reduced to 5 μg/kg because the daily intake of cadmium from drinking water by average adults will run up to 20 μg when they take 2 L of drinking water daily (contents of cadmium is 0.01 mg/L). This amount of cadmium is almost one-third the recommended ADI for cadmium. On the other hand, as it is so difficult in reducing the amount of cadmium contained in food that the Working Group considered the limit of cadmium in drinking water should be reduced as low as possible.

This suggestion is a very serious blow to us. Now we are planning to make a nationwide survey on amounts of cadmium in various foodstuffs, including our staple food, rice, which is thought to contain small amounts of cadmium to some extent; and on the other hand, we have to deliberate the improvement of the water purification system in order to fulfill the Working Group's recommendation.

The second problem: regarding the upper limit of selenium in drinking water; at present we have no control on its limit, but U.S.A. and WHO have its limit of 0.01 ppm.

In our country, our industries dealing with selenium have also made great development recently, and as the result, river water pollution with various factories' drainage containing selenium has become a serious problem for environmental safety.

The Ministry of Health and Welfare and the Environmental Agency in Japan have started investigations on the quality of river water, which tends to be affected by selenium compounds. Our Institute has also begun evaluating the toxicity of selenium from the viewpoint of its metabolism, using isotope-labeled selenium compounds and pharmacological long-term feeding test data, taking into consideration Dr. K.W. Frank's report (J. Nutrit., 8, 615, 1934) on horse disease caused by eating grass containing selenium, or Dr. M.I. Smith's report (Public Hlth. Rep., 52, 1171, 1937) on selènium poisoning among inhabitants living in the same region, and so on.

So far, we have observed a decrease in growth rate, inhibition of food intake through the examination of the subacute toxicity of sodium selenate by continuous oral administration of 1 and 5 mg/kg/day for 30 days in male and female rats. On the other hand, it has also been discussed, an enlargement of spleen and kidney, edema of pancreas, degeneration of liver cells, and subacute yellow liver atrophy, as outstanding pathological lesions, mostly in female rats.

The third problem: In March of 1974, five persons living in a town located in the southwestern part of Japan suffered from severe paralysis of legs and arms, with difficulty in walking. After the hospitalization, it was clarified that they were poisoned with acrylamide contained in their drinking water coming from a well. Later, it was learned that a ground engineering work was being carried out around that time and a great amount of acrylamide solution was used for the purpose of hardening the underground earth in the vicinity. Therefore, it has been supposed that the acrylamide had gradually reached to the underground water and contaminated it by being forced through the earth. Consequently, the contaminated drinking water caused poisoning to the humans. The Ministry of Health and Welfare has made a very serious issue of the incident, and our Institute started to urgently establish an analytical method for testing acrylamide in drinking water by adopting the procedure of distillation under low pressure and gas chromatrography with flame ionization detector. The detectable limit is 0.1 ppm.

Thereafter, the Ministry of Health and Welfare has issued a regulation that acrylamide in drinking water could not be detected by this method.

The fourth problem: recently Dr. J.M. Hill and Dr. G.M. Hawksworth have clearly published the following facts: the drinking water supplied to Worksop (a town in Great Britain) contained over 90 ppm nitrate. According to the Japanese drinking water quality standards, the concentration of nitrate (NO^-_3) is less than 10 ppm. Their analysis of the cancer deaths by site, age and sex, compared with a number of other towns, revealed that the number of deaths from stomach cancer in Worksop was unexpectedly high. The death rate of males was 32% higher and the death rate of females was 62% higher than those reported from the other towns. According to this significant information, we have to pay attention to the concentration of nitrate in drinking water from the viewpoint of the formation, in the human body, of nitrosamine by both nitrite from nitrate and secondary amines.

We cannot expect a rapid improvement in the near future of the environmental pollution in the human society, but I think that we have to pay particular attention to specific chemical contaminants likely to be contained in drinking water.

CRITERIA AND PROGRAMS FOR DRINKING WATER QUALITY IN CANADA

Emmanuel Somers

In considering the development of this topic from the viewpoint of our Federal Department, it is necessary to first consider ambient water quality standards and criteria - for this is the source of drinking water quality. Where fresh water is concerned, Canada has been blessed with an abundance: some 7.6% of the land area is covered by water. With much less than 1% of the world population, Canada contains 9% of the total flow of all rivers in the world. This has enabled us to provide the vast majority of the population with running water, most of which is piped from central sources such as municipal treatment plants. The growth in the number of municipal water systems since 1940 has been considerable - from 6.5 million to 14.3 million people served in 32 years (1).

And yet, six years ago, our Federal Parliament passed The Canada Water Act, the preamble of which reads: "...pollution of the water resources of Canada is a significant and rapidly increasing threat to the health, well-being and prosperity of the people of Canada and to the quality of the Canadian environment at large, and as a result, it has become a matter of urgent national concern that measures be taken to provide for water quality management in those areas of Canada most critically affected."

Threats to the quality of our water are common to many parts of the world - a head-long drive for industrial growth with a concurrent neglect of environmental effects. What I should like to do is discuss Canada's response to the ever-increasing need for quality water, in light of current disruptions of aquatic environments.

We, of course, recognize that in any industrialized country, the degradation of water cannot be totally reversed. It would be impossible to restore all Canadian waters to their original, pristine condition. Particularly in view of the amounts of water Canadians use daily - in excess of 31 billion imperial gallons per day are withdrawn for all uses (1). This amounts to 1400 gallons per person per day, used for: public supply (municipal and rural), industry (manufacturing and mineral industries), agriculture, and thermoelectric generation.

We can, however, plan the future use of this resource, on a restricted basis in many cases, around the anticipated economic potential and technological demands on the water. It would be worse than useless to promulgate a national standard for surface water quality in a country the size of Canada.

The variability in present quality of Canadian surface water, together with the multiplicity of uses, would require an almost case-by-case examination. A national standard, while perhaps appropriate to many water bodies, can lead to the pollution of high quality waters, as well as placing impossible demands on still other waters which would not meet the standard even in their natural states.

Water quality should not be modified greatly in any given characteristic, sufficient to fall outside of the prescribed criteria for a specific use. The long-term change may be quite significant, and the way in which it affects the aquatic populations may have devastating consequences. For example, the threat of oil spills is a serious potential problem, particularly in the Great Lakes and Arctic. The low temperatures and the presence of ice in Arctic waters for much of the year pose problems which, so far, are unresolved. Experience in combating the 1970 "Arrow" oil spill in Chedabucto Bay, Nova Scotia, where low temperatures and ice conditions prevailed for part of the time, suggests that consequences of oil pollution in the Arctic could be extremely serious. In the Chedabucto Bay spill, 7000 seabirds and some seals were killed and a commercial fishing area was closed.

We should at this stage define our terms, in particular, those concerned with criteria and standards (2):

Criteria - are scientific requirements on which a decision or judgment may be based, concerning the suitability of water quality for the preservation of the aquatic environment, and/or to support designated uses. Criteria include expressions of the dose/response relationship in terms of the effects that are known or expected to occur whenever or wherever a detrimental factor and/or pollutant reaches or exceeds a specific level for a specific time.

Standards - evolve from established criteria and are legally prescribed units of pollution and/or deterioration, that are established under statutory authority. The types of limits that have been adopted for drinking water are: objective, acceptable, and maximum permissible (3). Essentially, these may be defined as follows:

Objective - long-term goal representing water of the very highest quality.

Acceptable limit - limit which, if exceeded, would be objectionable or deleterious to human health.

Maximum permissible limit - fixed standard for substance known, or suspected, to be linked to human health.

Particular standards and objectives should reasonably reflect the present and future uses of a water body. Recently, however, the use of the three levels of limitations has come under review. Maximum permissible limits and objective limits are the extremes of the water supply quality spectrum, with the Objective Limit as the endpoint to which we constantly direct our efforts. It is felt that Acceptable, while acting as an index to raw water suitability within the context of the definition, is superfluous as a limit and will probably be dropped. The principle of multiple use of water has become enshrined in progressive industrialized societies where it becomes necessary to provide ample water supplies for a wide range of purposes, from domestic consumption to industrial waste disposal. However, this complicates enormously the establishment of water-use priorities. The most satisfactory solution would appear to be the establishment of criteria in several water-use categories; e.g. agricultural water, shellfish culture, industrial, recreational, the most important of which is, of course, water sources for drinking water supplies.

Drinking Water

Where drinking water is concerned, there is a definite relationship between the quality of the source and that of the treated water. Three types of treatment processes and modifications are used in Canada:

- Type 1: Disinfection is intended to improve raw water biological quality. Chlorine is commonly used.
- Type 2: Chemical Coagulation - Sedimentation - Filtration - and Disinfection. This process of chemically coagulating impurities so that they will settle or be filtered out is commonly referred to as the conventional, or complete, treatment.
- Type 3: This is an application of conventional treatment with additional processes; for example, softening the water (using ion-exchange and other methods, taste-odour removal, etc.).

Where the source of community water is heavily polluted, only the use of the most complete treatment technology will produce water safe for human consumption. Chlorination, used extensively in many communities in Canada, does little to reduce objectionable colour, taste, odour, and turbidity of the raw water source.

With the multiplicity of uses of water resources and the increasing threat of pollution, it comes as no surprise that several areas of government are involved in establishing criteria and recommending standards. In Canada, the quality of ambient water is the concern of the Department of the Environment at the Federal level. Drinking water, as it relates to the health effects on humans, is an important concern of the Department of National Health and Welfare. The prime responsibility for health in Canada rests with the Provinces, so that standards for drinking water are not set by the Federal Government. However, guidelines have been developed by joint Federal-Provincial consultation. The Federal Government uses its resources to establish criteria and guidelines for water quality and, in some cases, of ambient water pollution, it may enter into agreements with the Provinces to mount programmes for the control of pollution of specific water bodies. As occurs with other countries which have Federal systems of Government, we in Canada sometimes have jurisdictional difficulties in coordinating our efforts toward the same ends. There are some twelve Federal Acts pertaining to water quality, which have been passed by our Federal Parliament over the last 100 years. The most important of these are The Canada Water Act and The Fisheries Act. Although some of the more recent ones are more comprehensive in dealing with quality of water (e.g. The Canada Water Act), one can appreciate the complexity of a legislative approach controlling a substance with such varied uses. Water must be fit to support fishing and navigation, and must be fit as food for humans.

The first Drinking Water Standards were developed and approved by the Federal Government in 1923 for use aboard shipping vessels on the Great Lakes and Inland Waters. The Standards concerned bacteriological quality of water used for drinking and culinary purposes. Since that time, Canadian authorities have periodically reviewed, borrowed, and applied USA and WHO Drinking Water Quality Standards. It was in 1968 that the first Canadian Drinking Water Document was brought out by a Federally-convened Committee with representatives of the Canadian Public Health Association. The resulting publication (3) contains recommendations for safe limits and desirable sampling frequencies for toxic inorganic and organic chemicals (including pesticide residues and toxic metabolites of aquatic microorganisms), microbiological hazards, radioactive materials and gross radioactivity. The Standards also deal with the more subjectively measured physical characteristics: taste, odour, and colour, as well as turbidity and temperature. Objectives, acceptable limits and maximum permissible limits were set for coliform organisms and pesticides in both raw and drinking water, the latter accompanied by recommendations for the level of treatment required to meet the Standards.

The basis on which the limits were set, i.e. the criteria, are contained in this document. These criteria largely emanated from the United States Public Health Service Drinking Water Standards of 1962, but the Committee, in its own words "took note of the changing environmental conditions in Canada, their effect on drinking water supplies, the economic problems of treatment, sampling and analysis of smaller water supplies and the lack of judicious criteria, in some cases, for developing precise concentrations limits" (3). The effect on establishing standards of the so-called "changing environmental conditions" factor is shown in the case of fluoride, the recommended concentration of which is 1.2 mg/l treated water. In Canada's unique Arctic and Sub-Arctic zones which support a significant population, the recommended level was adjusted upward to 1.4 mg/l treated water to account for the naturally lesser inclination to drink water. This again emphasizes the need to consider criteria, out of which evolve recommendations in the context of the environment in which the population lives. The health effects associated with the quality of drinking water can be considered in terms of pathogenic organisms, algae and fungi, chemicals, radioactive substances, and interference with water treatment processes. There is today, somewhat of a preoccupation with the toxic effects of chemicals, since more and more, new organic pollutants and metals - the removal of which is difficult and expensive - are appearing in our ambient water from industrial and agricultural sources. In Canada, our problems do not involve only disease, but also water quality monitoring, obsolescence of standards and absence of a good statistical base required to attack present problems of water quality and water-borne diseases.

Deficiencies in Disease Statistics

Although incidents of acute waterborne disease outbreaks are reported in Canada, no national tabulation is maintained. We have no information concerning the comprehensiveness of the present reporting system. Until very recently, only the primary cause of death has been tabulated in mortality records: thus, chronic disease studies associated with water are difficult at best.

Deficiencies in National Information on Drinking Water Quality

The Federal-Provincial Working Group on Drinking Water was formed in 1974, to develop a mechanism for the collection of even general information on health-related aspects of national drinking water quality. Through this group, a modest beginning has been made in the collection of general information such as the nature and degree of chemical, bacteriological, and radiochemical analysis performed across the country.

Obsolescence of Present (1968) Drinking Water Standards

During the past few years, it has become evident that the 1968 "Canadian Drinking Water Standards and Objectives" (3) require substantial revision. For example, there is no standard for mercury, the radiological standards are probably at too high levels as are the levels for pesticides. The Federal-

Provincial Working Group on Drinking Water is presently reviewing the 1968 Standards, and hopes to produce new ones by 1977.

Inadequacies in Monitoring

Since the importance, nature, and frequency of various kinds of monitoring are a function of the various constituents of drinking water, we may subdivide this area according to constituents types.

Virological and bacteriological monitoring

Although substantial efforts are being made in the United States and elsewhere, there are no official methods to isolate, culture, or identify viruses in drinking water at this time. In the absence of official procedures for virological testing, no routine monitoring of drinking water for viruses is undertaken in Canada. Of the several treatment methods which have been suggested for virus inactivation, the most reliable one is the elevation of pH to 10-11 for a short time, followed by pH re-adjustment to an acceptable value. The bacteriological quality of drinking water is defined in terms of the number of coliform organisms detected during specified sampling periods by specified analytical methods (3). In most Canadian jurisdictions, the number of analyses required varies from two per month for communities of 2,000 people or less, to 100-400 per month for communities ranging in size from 100,000 to 2,000,000 people.

Although the number of coliform bacteria present in a given sample of natural water by no means indicates the total population of pathogenic bacteria, experience has shown that the number of coliform organisms present after disinfection is a reliable indicator of the effectiveness of the disinfection process. Furthermore, since most waterborne disease outbreaks result from the contamination of drinking water by sewage wastes, drinking water standards specify also that raw water (water to be processed by a water purification plant) is periodically tested for fecal coliforms; the presence of fecal coliforms is regarded as evidence of recent contamination of the source by sewage. In spite of increasing recent criticisms of standards based on coliform indicators, it is still true that no cases of bacterial waterborne disease have been observed in people who consume water classified as safe by the coliform indicator criteria.

Because of historical precedent, the obvious nature of acute bacterially-induced outbreaks, and the general level of public knowledge about water quality, bacteriological monitoring is generally the most rigorously-enforced and best-funded monitoring activity. This situation is, in itself, a matter of concern to us, since it engenders ignorance of the importance of chronic disease studies, toxic chemicals, etc., associated with drinking water.

Monitoring of physical properties

Colour, odour, taste, turbidity, temperature and pH are the physical properties of importance to water quality and limits have been set in the 1968 Standards (3). The recommended practice is that these parameters be determined on a weekly basis for both raw and treated water, and at least seasonally from representative points within the distribution system.

Colour, odour, taste and turbidity may, however, also be indicative of the presence of undesirable pollutants in a water supply, and high levels of turbidity decrease the effectiveness of chlorination dramatically. From the public health point of view, physical properties are included in standards mainly to ensure the aesthetic acceptance of a given water supply.

Toxic chemicals

This group of drinking water constituents includes inorganic elements (such as arsenic, lead, selenium, etc.), known organic compounds (such as Aldrin, DDT, phenol, etc.), and a host of unidentified organic and organometallic substances. Current drinking water standards specify semi-annual analysis, except for the pesticides, which are to be analyzed quarterly; provision is also made for less- or more-frequent analysis depending upon local circumstances.

Drinking water standards, including the 1968 Canadian Drinking Water Standards, include a test procedure for estimating the dissolved organic fraction of drinking water by an adsorption-extraction-weighing sequence of the organic material which can be adsorbed on activated carbon. Development studies on an improved version of this test are in progress in the Environmental Health Directorate and it is hoped to introduce the test on a widespread basis in Canada if the development work indicates that the improved method is satisfactory. Toxicological studies on the organic extracts obtained from this test procedure are also being considered, as is the possible utilization of the technique as a general qualitative screening measure.

Radiological monitoring

Radiological standards specify limits for individual radionuclides and/or for gross radioactivity of a composited drinking water sample, taken on an annual basis: specialized analytical facilities are required to undertake the requisite analyses. Radiochemical monitoring of water is carried out in five areas of the country; all of these analyses are performed on raw water.

Arsenic

The Maximum Permissible Limit of arsenic is 0.05 mg/1 in Canada's 1968 Drinking Water Standards. Earlier this year, levels of 0.07-0.08 mg/1 were discovered in water in British Columbia used only occasionally for drinking. The World Health Organization reported that similar levels have been found in some drinking water supplies in Latin America.

Nitrilotriacetic acid (NTA)

It became apparent in the late 1960's that sodium tripolyphosphate, widely used as a chelating agent in detergents, was contributing to the eutrophication process in lakes and rivers. Federal legislation in 1970 required reduction of phosphates in detergents, and the industry responded by replacing it with another detergent builder - nitrilotriacetic acid. By 1974, 23 laundry detergents contained up to 10% NTA trisodium salt and current estimates indicate that up to 50 million pounds of NTA are used in Canada each year. Although we concluded, and continue to believe, that the existence of NTA in the Canadian environment is unlikely to produce adverse health effects, we were concerned about its widespread unrestricted use, consequently, monitoring programs with particular emphasis on drinking water were established by the Department of the Environment in 1971. Analyses of most treated water samples showed a content of NTA below or at the limit of detection (which was 10 ppb using a polarographic method) and nine exceeded 40 ppb NTA.

The principal health concern derives from the fact that NTA, when administered in the diet in massive doses to mice and rats, may produce carcinoma in the urinary tract. However, as there is a factor in excess of 50,000 between the highest levels of NTA found in Canadian drinking water and the no effect level for carcinogenicity in the most sensitive species of animal tested, we consider that no health hazard is posed by the present use of NTA in Canada.

Organochlorine compounds

We are now conducting research into the identification of organochlorines in drinking water. Pesticides are in wide use in Canada and many are well-known for their toxic potential. Two of these, chlordane and lindane, have been detected in tap water in sufficient amounts to warrant investigation to determine the presence of other pesticides and seasonal variation of contamination.

The recent report of the United States Environmental Protection Agency, disclosing the existence of trace amounts of carcinogenic compounds in drinking water, suggests that even more extensive chemical analysis of drinking water is required, although extrapolation of the possible implications to human health is as yet only tentative.

Of great concern, also, is the recent discovery of chloroform in small amounts in the drinking water supplies of many American cities and several Canadian cities. Chloroform is known to be a carcinogen when administered to mice in very high doses. We are proposing surveys of volatile organohalides in the drinking water supplies of Canadian cities.

Water treatment methodology

In general, very few technological advances have been made in water treatment plants in Canada during the past 50 years. The most advanced treatment plants generally rely on a combination of flocculation,filtration, and chlorine disinfection for water purification. Dissolved substances tend to pass through this form of treatment with only minor changes in their concentrations.

Potential technical developments are carbon adsorption, which reduces the concentrations of both dissolved organic and inorganic substances, and the use of ozone as an alternative to chlorine for the initial disinfection of drinking water. A few carbon adsorption treatment plants are in operation in Manitoba.

Ozonization is in use in Canada in several locations in the Province of Quebec. Although ozone is effective in deactivation of viruses, chlorination is preferred because it also confers protection against bacteriological contamination after water purification. Because there seems to be little information available on the gross effects of ozonization on the chemical characteristics of dissolved organic substances, a collaborative research project with the Canada Centre for Inland Waters was initiated last year.

Other areas of concern related to water treatment methodology include our lack of knowledge about water storage facilities and provincial procedures for use in case of emergencies involving drinking water. If insufficient storage capacity is available in a water treatment plant, untreated water could be used in the distribution lines in order to maintain pressure during process breakdowns or repairs to the treatment plant facilities. We suspect that many treatment plants, lacking sufficient water storage capacity, would be tempted to follow this procedure.

Regional problems

Certain areas of Canada are presently subjected to high concentrations of hazardous and potentially hazardous substances in their drinking water. Examples include asbestos, nickel, petroleum oils, mercury, and arsenic contamination of raw water.

Programs

I have described the mechanism by which we in Canada establish and review drinking water criteria and standards. The 1968 Standards were established by a committee comprised of representatives from the Canadian Public Health Association, the Federal Department of National Health, several universities, and water pollution and water resource control agencies.

The process of review of criteria is now an on-going one, with the establishment of a Federal-Provincial Working Group on Drinking Water. The reasons for the review I have already mentioned. Our method, one of cooperation at both levels of Government, is a sound one, I believe. It is difficult enough to obtain the necessary scientific information and to act on it to establish standards, without having to conduct arguments on jurisdiction between levels of Government. A joint committee, involving expert representatives from all Governments at every stage, can expedite the process of information gathering, as well as avoiding unilateral and uninformed regulation.

The Department of National Health and Welfare is presently considering a course of action to rectify the inadequacies in data-gathering and analytical methods which I have mentioned; it is hoped that in the near future the Drinking Water Standards and Objectives will be revised and that guidelines on contaminants, whose toxic potential has only recently been realized, for example mercury and some pesticides, will be included.

It has been proposed that we establish a National Data Base of information on water quality from both Federal and Provincial monitoring and surveillance programmes on drinking water, and that analytical methods be developed to be used in these programs to provide the basis for more meaningful standards. In addition, we hope to collect for analysis, disease statistics to detect trends associated with water quality and to undertake epidemiological studies on the relationship between disease and specific contaminants.

The outcome of this course of action would be two-fold: first, the establishment of sound standards and objectives reflecting the most up-to-date comprehensive information on contaminants to which the public is now exposed, and second, the support of existing Provincial programmes, leaving ultimate regulatory control in the hands of the Provincial Governments.

The Federal Department of the Environment has developed

a National Water Quality Data Bank. This system is designed to accept chemical, physical, bacteriological, biological, and hydrological data relevent to water quality for surface waters, groundwaters, wastewaters and sediments. A network of collection and sampling stations extends across the country. The stations are operated by Federal Government and Municipal employees and samples taken are tested for major ions and physical characteristics.

Conclusion

We consider that there are five separate courses of action necessary to promote and maintain the potability of water used for domestic consumption:

1. The revision of Canadian Drinking Water Standards and Objectives in the light of current knowledge and concerns.
2. The creation of a National Data Base of information on water quality from Federal and Provincial monitoring and surveillance programs and the analysis of the data.
3. The development of analytical methods to be used in Federal and Provincial monitoring and surveillance programs, and to provide the basis for more meaningful standards.
4. Initiation of toxicological studies on contaminants of drinking water such as organochlorines. In addition, we hope to undertake mutagenicity and carcinogenicity studies on organochlorines identified by your EPA and not at present under study. We plan to undertake interaction studies of environmental contaminants.
5. Analysis of disease statistics to detect trends associated with water quality and initiation of epidemiological studies on the relationship between disease and specific contaminants, including asbestos and chlorinated compounds.

As a result of the problems shared by our two countries in the area of water pollution and health protection, a high degree of rapport has developed between scientists working in both Federal Governments. In view of the certain longevity of many of these problems, I am sure you will agree that this rapport should be maintained and strengthened at every turn. ultimately for the mutual benefit of future generations of Canadians and Americans.

REFERENCES

(1) "Canada Water Year Book 1975", 232 pp., Information Canada, Ottawa (1975).

(2) Working Group on Water Quality Criteria of the Sub-Committee on Water Quality, Interdepartmental Committee on Water, Department of the Environment. Guidelines for Water Quality Objectives and Standards. Technical Bulletin No. 67, Department of the Environment 1970.

(3) "Canadian Drinking Water Standards and Objectives 1968". Queen's Printer, Ottawa (1969).

DISCUSSION

DR. MRAK: I think this was a very interesting and exciting sesstion this morning. It is the first one where much was said about microbiology. How about others of you, regarding viruses and microbes?

DR. SMEETS: I do not know if I have to worry about getting sick from traveling. The only point, and that is what I mentioned in my paper, seems to be the lack of a systematic monitoring of microbiological organisms in drinking water in the Western European countries. As Dr. Somers has mentioned, the problem involved is this, and we realize that there are quite a number of problems from the analytical point of view, and I mentioned also in my paper that we tried in Lyons this year to bring the best qualified experts together, to try to find methods to harmonize the analytical techniques in Europe. You may be interested, since you mentioned it, Dr. Somers, that I had a discussion with some EPA people last week. They were wondering whether the deep freeze technique could be used for stabilizing the samples. And I learned that EPA is developing this technique to send the samples to the various laboratories. This is an interesting point, and we will hear more about it in the near future. This might be an interesting suggestion for you, too.

MR. DONALDSON: In Latin America, as far as the viruses go, we at the present time have little or no data on this problem, so I cannot make too much of a comment concerning it.

DR. KAWASHIRO: There are many problems regarding water quality standards. I would like to mention some other kinds of programs regarding the drinking water quality in Japan.

We have the same problem regarding chromium. Inorganic chromium and organic chromium is a very serious problem in my country.

DR. MRAK: Mr. Ted Schad, Executive Secretary of a panel of National Resources of the National Research Council, National Academy of Sciences, is here. His whole life is concerned with water. He has been here throughout the meeting, and I thought he might like to comment at this point.

MR. SCHAD. It has been a great privilege to be here.

I have come to the water business more as an engineer, concerned with governmental policy rather than as an expert. But, I would like to comment on the last point, which I think is a most significant point that Dr. Somers made. We are spending all of this effort and time to purify and produce a water product to meet one percent, or one half of one percent, of the water that really is needed to be this pure. I think his estimate is probably too high, particularly when you relate it to the entire water supply of the municipalities. I just don't know if many of us drink that one and a half gallons. What has happened, at least in this country, and probably not so much in the other countries, is that our water industry has gotten so efficient that on every occasion, when I have had time to look into the economics of providing a dual water system, where we have a drinking water system and water for other purposes that don't need to be that pure, it has come out ahead to provide one system and purify all the water. The most recent situation with which I was associated was the water supply for the new town of Columbia. This is a new, small city, eventually to reach about 100,000, between here and Baltimore. A very thorough study was made in this case, because some people started out with the idea that here is a chance to try something new and put in a dual water system, where we only have our drinking water purified and the other would be to a lesser standard. There was just no comparison economically between the figures. It was cheaper to have one system and treat the whole water supply.

Now, that was about six or eight years ago that that study was finished. Since then, there have been lots of changes, and we now have the Safe Drinking Water Act, which brings the full power of the Federal Government into the drinking water business, even though the law says that the responsibility is still with the states. This is a favorite ploy in this country; when you start out with an Act that says that the Federal Government hereby recognizes the rights and responsibilities of the states, and then goes on to set standards. Then, there is a lot of fine print in the Act! Really, the Federal Government is moving into the water supply business in a big way, through that Act, and this may very well, if those standards are set in such a way that requires a lot more treatment, tip the economic balance toward the dual water system. And I am waiting with some interest to see just what comes out the next time somebody figures the economics of a dual system. Because it just doesn't seem efficient to purify that extra 99% of the water. I guess when you look at the whole thing it might be 5% drinking and 95% general use water. The other option, of course, which used to be more prevalent than it is now, is in foreign countries where you just don't drink the water, to buy bottled water. I suppose we are going more and more toward bottled water in this country, now, and this may bring about a change. I think this is one of the significant changes that I see ahead in the drinking water standards.

I think a lot of the other discussion here, and particularly yesterday, on the carcinogenicity of various chemicals in water, raises the question of how do you make the decisions as to safety when you don't know the answers. This is a field that intrigues me, now, but it is really not my field. But, we do have a major effort at the Academy looking into this subject of decision making on the basis of inadequate toxicologic information. We are really looking at the whole business of how you use scientific and technical information in making governmental decisions. It is a difficult subject, and we may end up falling flat on our face with it, because we are trying to bring the scientists and the economists and the political scientists and some politicians together, and try to evolve a process or a procedure as to how you answer these questions. This discussion was interesting to me yesterday, and I wish it could have gone on a few more hours, because some day, we are going to come out of this with some kind of an idea of how to deal with these problems, where there aren't any real scientific answers. This is where most of my effort is devoted right now. And from that viewpoint, I am very interested to be here, and I thank you very much for inviting me.

DR. TRUHAUT: I have listened to you with great attention. And, of course, as hygienists and toxicologists, we are very often in a bad position. But we, and the politicians, have to realize that the public must know that a scientist can be defined as a man who selected a certain kind of ignorance, or as a man who has some gaps in his ignorance. For this reason, we have to rely, in my view in many instances, on what I would call "value judgment" and "good sense." Otherwise, we will not succeed. We are not in a Paradise, we are on Earth. Here we have to face pragmatical problems and try to solve them, I repeat, with good sense.

DR. COULSTON: I am intrigued, and I will relate to the program, Mr. Chairman. I would like to mention the question of NTA (nitrilotriacetic acid), along the lines mentioned by Mr. Schad. In America, we have struggled with this problem of NTA as a substitute for phosphates now for at least ten years. We still cannot make a decision, based on the problems of teratogenicity, whether NTA is an imminent carcinogenic hazard to man. You cannot prove that it is safe to use by man, you can not prove that it isn't. In defense of EPA, they are on the horns of a dilemma. But fortunately, I think, for science and for good sense and judgment, Canada bit the bullet, and decided they would use it as a replacement for phosphates rather than use extremely caustic chemicals in laundry powders. Maybe, after watching what is happening in Canada for ten years of NTA use, we could make up our minds in the USA. Now, why could Canada make a decision and why can't we? Why has most of Europe made the decision to use NTA, and why haven't we? This is very important!

DR. SOMERS: You are very flattering, Dr. Coulston. But, of course, the reason we made the decision wasn't quite so rational. What happened, as I understand it, was that a political decision was made that we should not use phosphates. And then, the Minister at that time, was convinced by his advisors that phosphates were an imminent hazard towards our particular form of lakes. He wanted, therefore, a replacement. And it wasn't really that we chose NTA, but that we moved away from phosphates, and picked NTA as a safe substitute. That was the background of that decision. I wouldn't ever claim that there is any more sanity north of the 49th Parallel than south.

DR. ERNST: Concerning Canada and NTA, there are lakes around this world which definitely are not phosphorous, but nitrogen limited. Probably some of them may be in Canada, too. Has there been research on what is in different lakes? And, if not, does the government expect that there are so few lakes limited in nitrogen instead of phosphorous that it is just not worthwhile considering it? Is there any danger that the elements which are limiting the biologic production in a lake may change by increasing pollution levels?

DR. SOMERS: I do not know. I do not have the answers to all the questions you have asked. But, I suspect that as the Department of Environment is still satisfied with the use of NTA, that in fact, we are not in a situation of nitrogen limitation in our lakes. Phosphate must be the one concern. As a matter of fact, the gentleman here from Procter and Gamble, may want to give a more authoritative answer.

MR. KRUMREI: To answer Dr. Ernst's questions, one of the difficult problems is to try to determine what is a limiting nutrient in a lake. There are a lot of limnologists who maintain that phosphate is the limiting nutrient. This is based on laboratory tests in glassware. To a large extent, there are lakes that are probably limited in phosphate, and so we cannot argue with the Canadian government in its wisdom in passing that law. However, based on what they have seen so far, there do not seem to be enough lakes limiting in nitrogen, and particularly with the levels of nitrogen which are going in as a result of the use of NTA, to cause any major concern.

We do not advocate a general passage of regulations

forbidding the use of phosphate, without very carefully studying where all the phosphate is coming from. In most countries, the usage of detergents containing phosphate contributes only a portion of the total phosphate, and by elimination of the phosphates in detergents, you are eliminating that fraction. But, there is still a lot of phosphate coming in from human waste, from industrial waste, agricultural runoff, and so forth, which are not controled, unless you remove phosphate at the sewage treatment plant. By and large, Canada has been doing quite a bit of that, as well, and the United States is going in that direction. In the agreement within the International Joint Commission between Canada and the United States, all of the Great Lake States are moving to remove phosphate from the sewage, which will take care of detergents as well as human wastes.

One more point, Mr. Chairman. Canada had the choice, really, when they requested that we lower phosphates in detergents, of using a material such as NTA, which is an excellent builder, which has been proven to be safe in all applications, with the exception of the very high level feeding studies, probably through a different metabolic route which establishes those cancers, than through the normal low level ingestion. They had the choice of using NTA or other materials, which were quite caustic in detergents, as Dr. Coulston mentioned. Sodium carbonate, and others. The United States took the route, initially, of going with some caustic detergents. There were some unfortunate incidents in the case of child ingestion, and Canada decided they did not want to duplicate that. So, they worked very carefully on the safety of NTA, and came out in favor of that.

DR. MURPHY: Dr. Somer, I was interested in the logic that you presented for your decision to allow the use of NTA, and you cited a specific figure, which I will risk calling a safety factor, between your experimental animal studies and the maximum predicted ingestion of man, of 100,000. I wonder if you apply this principle, or any kind of safety factor consideration, to other chemicals in water, and what you do with so-called experimental tumorgens in foods. Of course, in the USA, we have the Delaney Clause that has caused a lot of controversy, and set precedents. I don't know that Canada has a similar restriction.

DR. SOMERS: No, we do not have a Delaney Clause. But, you know, though you have a Delaney Clause, you have aflatoxin in your foodstuffs, and you have nitrosamines in your bacon. And you have to bring in regulation and control, in spite of whatever particular law may be used on this question.

So, I think what we do, and I don't want to make any great claims in this debate between the USA and Canada, is to try and make an assessment. Sometimes we can do it in terms of a factor, as I think we can with NTA, because we know the maximum level of NTA that could occur in drinking water. And sometimes, we can come out with a safety factor, like 100,000. And then, you know, you can use mental arithmetic, and you can say that that means there is a chance of one in fifty million of someone developing a cancer.

I think that is the sort of calculation that you have to do with the public. Not that anything is absolutely safe, but that there is a particular, and all things being equal, we want to still continue to eat peanuts, and we have a level of approximately 15 ppb, which is our guideline. But essentially, has all the stages of a tolerance. In the States, though you have the Delaney Clause, you still have about 15 ppb, surely, of aflatoxin in your food.

DR. MURPHY: Of course, this is where these issues come to such controversy, because the Delaney Clause begins to be interpreted and considered with these incidental contaminations, which are a different kind of a situation from direct food additives.

DR. COULSTON: I have a related point to what has just been said. Let me say this, very strongly; although the FDA has a Delaney Clause, they have rarely used it, and never on an important issue. It was not used for cyclamate. It was never used for the herbicide in cranberries. They have the authority to restrict or ban chemicals from food without using the Delaney Clause. The Clause specifically says that the experiments have to be good ones, acceptable to the Commissioner. I think we use the concept of the Delaney Clause too much to "scare" people. I want to make it clear that the Delaney Clause has never been used on a major issue by the FDA. They don't have to. Perhaps, they may try to use it on the diethylstilbestrol (DES) because Congress is making an issue. Now that is a different twist. Congress, by edict, by law, may force them to use it. So, let's not throw Delaney Clause tactics around too frequently. There is a lot of leeway for the Commissioner to make decisions, as Dr. Somers has remarked. He doesn't need to employ the Clause. We do have nitrosamines, we do have aflatoxins and many other chemicals in our foods.

DR. SMEETS: I would like to make some comments and ask some questions. We have one point to keep in mind all the time. We speak, today particularly, about the criteria of drinking water. Now, again, that is only one, and perhaps not the major intake of chemicals, by ingestion, by man. We have to remember that the other, and perhaps more significant, one is food. Chemicals in air would contribute a small part to the whole body burden of the general population. I mentioned that yesterday, and I would like to emphasize it again. There is no sense to spending huge amounts of money to remove a certain chemical when you get it in food in much greater amounts.

In this connection, it is interesting to say that spinach, for example, has a very great concentration of nitrosamine compounds. I just learned this last month. It is interesting to say that babies usually, and little kids, don't like to eat spinach, and maybe they have a natural resentment, which turns out to be a very healthy one.

But again, I would like to say that nitrate may change somewhere along the line to nitrosamine compounds. The suggestion was made about bottled water, at a recent meeting in Copenhagen, on nitrogen compounds in water.

Again, we had the suggestion of drinking more bottled water, when we discussed the dual systems. Now, I think we have to rely on the cost of a dual system, the cost evaluation of a dual system, in comparison to a single system. I still think that it will probably take a long time before a dual system will become economical. But is bottled water really the solution? It may be the solution in some countries. France has very well adapted itself to bottled water. I doubt that the French population at large would drink very much tap water. But the world is not composed only of French people, and bottled water has been proposed during the discussion of nitrogen compounds in water to solve the problem of safety for babies up to six months. In principle, the water system has acceptable nitrate concentrations as high as 100. That is what the European standard did at the time, simply because countries do use river water with this kind of concentration and no problems have been observed. But we do have a world population to be considered, as a matter of fact, we are concerned with world population, and I don't think that bottled water will be a solution for world population in a large while. Take India! Are you going to supply the Indian people with bottled water? I doubt it, at least not in the near future.

I think our WHO standards has made the point that sampling should be done at the end of the system, because that is where you use it. If you take the sample at the outlet of the water treatment plant, you probably are well off, but a lot can happen in the pipes. Therefore, what counts is the water that comes out of the taps. Perhaps sampling at certain points in the pipes would serve a purpose, that may be so, but that would be good only to advise the engineers as to what care to take. I don't know whether that would satisfy us in setting drinking water standards.

I think flow rights is a particularly important point, since it has a number of issues. One of them, and I mention this perhaps to get a response from some of our speakers, is, of course, that you would like to change the concentration of fluorides in water in accordance with temperature. This idea was really first introduced by the US Public Health Service in 1962. It has been adopted by the WHO for European and International Standards, with conversion to the metric system.

Now as far as the Common Market is concerned, I noticed through a little personal communication, that they did not use the lowest required guide level, which means really, the recommended level. Now fluorides are an issue which becomes political. The WHO, as such, as come out with a resolution by the World Health Assembly a few years back, supporting the fluoridation of water. In the British Islands, where a lot of tea is consumed, you must consider a reduced concentration, because there is a lot of fluoride in the tea, and so on. But the USA still has a problem, too, because of certain areas where local population, or rather local citizen groups, are fighting fluoridation as an addition to natural water. They often believe that everything should be natural, as given by Nature, or by God. Well, again, that is an issue which I think the Health people, administrators, politicians, and those who make regulations for water standards, will have to take into consideration. I would like to make this connection, that Sweden has now by law abandoned, or forbidden, the use of fluoride simply because of the drive of one parliamentarian against it, which somehow managed to pass the majority of the house. That, despite the fact that Sweden, all but one little community, has never used fluoridation in their water supplies. You can see, again, it became a rather emotional issue, rather than a practical one.

Now again, I said that the WHO Drinking Water Standards, both European and International, has introduced the polychlorinated hydrocarbons, with great reservations. But at least they faced them, or brought the issue in and included them into the standards.

MR. DONALDSON: You asked the question of sampling. But you also made the point that a lot of things happen between the tap - between the pipe and the tap. And I think in Latin American we have to carry that further. A lot of things happen between the tap and the mouth, because oftentimes, water is carried in various containers and what-have-you. The basic standards that are applied in most of the countries are WHO standards, or better. But as I pointed out, in reality, depending on the country, you end up with three or four levels of application of these standards.

DR. KAWASHIRO: In Japan, we usually sample just one litre. This includes the bacteriological and chemical tests.

DR. SOMERS: I don't know whether I have anything very useful to add, except to draw attention to the fact that analyses in terms of guidelines are always carried out by minicipalities, before the water reaches the consumer. And so, if one is concerned about possible contamination in the piping system, then there is no way of checking that, the way the present analyses are carried out.

DR. HUTTON: In the bottled water industry, we have been very concerned about the sampling that is done by the public water system at the originating source, not at the end of the pipeline. And there have been a number of articles that have come out, principally in the "Washington Star" (D.C.), actually pointing out that this falls down. We know of one public supply system that tests at the end of their pipe, and then buys bottled water to serve people on the perimeter, because it is so bad in the pipeline. It is an extreme concern, and it should be considered.

DR. SMEETS: I think Dr. Somers mentioned the problem of biological analysis. Now, may I simply mention here that this problem will be at least making one step backwards, with the publication of the analytical methods, including virology. A standard biologic method for water is being worked out. The water virologists have said that they will trust average laboratories, as far as sampling and concentration of the sample, but for tissue culture, which is of course, more difficult and has to be done by special laboratories that can do it. In other words, we cannot yet expect average water laboratories really to get to the point of counting and identifying viruses.

DR. SOMERS: What viruses are you talking about, other than hepatitis?

DR. SMEETS: There will be viruses other than hepatitis. I don't want to say more, because I am not myself a virologist, and I would rather refer back to the document that I don't have here.

MR. DONALDSON: Being a person who has had hepatitis, it is a subject of a great deal of personal interest, but I think that I have to qualify the answer, such as I would have to qualify all my answers. I am dealing with a continent which has 300 million people, 27 countries, a range of problems that go over the full gamut. In Argentina, in the water supply of Buenos Aires, they are quite concerned about this problem. They are looking into it, the problem of viruses. They have a committee that has been in existence for about two years, and they have not been able to come to any particular conclusion. But, when you look at the problem in its broad scope, within the whole range of problems in Latin America, that is not one of first priority.

DR. SMEETS: We are faced with an interesting question, a general one, that we have different groups setting different or similar standards on national and international levels. Very often, the experts involved are the same. And I already pointed out that the same people are able to set sometimes slightly different standards under different umbrellas. I think this is an important point, and I just wanted to bring it forward.

DR. SOMERS: Could I just address that, because I think it is an important thing. You can have different standards because you have different resources available in the different countries. You can establish the international standards, but the people who have to enforce it are on a national basis, and oftentimes on a regional basis within that country. Concerning the standards, the technical people come in and say here is what it should be. Then, someone, often a politician, must ultimately make a decision and say that these are what is feasible at this moment in this country, and that decision will often change within a short time.

DR. COULSTON: I have been listening to Mr. Donaldson and the others talk about the microbiological hazards. Now, we use chlorine, and I don't understand why only chlorine and not other chemicals or methods. Chlorine is a very good antiseptic and will kill almost everything, except if there are a lot of solids present in the water, you've got to use more and more of it until you couldn't drink the water. But the point is, I don't know why you have such a big problem microbiologically with drinking water, when you are putting chlorine into the drinking water. Can you explain this to me, please?

MR. DONALDSON: The point you bring up is an interesting one. Let me give the answer in this way. The problem of chlorination as we see it in Latin America is more a logistical problem than anything else. I can buy a chlorinator. I can install a chlorinator, that is a one-time operation. But the supply of chlorine to that chlorinator is something that must go on day after day after day. Secondly, for example, in Central America, the purchase of chlorine is a foreign exchange item. In many of the developing countries, foreign exchange is the most difficult thing to get. Therefore, the system breaks down. Thirdly, you must transport the chlorine to the site in which it is going to be used. And in many countries, the transportation of the chlorine to these sites are over difficult terrain, through back roads, and so, they don't arrive on any particular consistent basis. The chlorine does work, but the mechanics of getting it there is another matter.

DR. MURPHY: Maybe in 50 or 100 years, Latin America will worry about chloroform resulting from the chlorination, as the USA is doing now. But I was interested in Mr. Donaldson's remark regarding the distribution of aid within the country for promoting improved water quality. And, as I recall, you had about an even distribution of urban/rural population, but a relatively or very small proportion, going rural. I wonder if this is a case where you can handle the rural problems less expensively, and yet deliver the necessary education. How do you reach the conclusion to distribute the aid in that proportion?

MR. DONALDSON: You have hit a subject that is very dear to my heart. But I should clarify one thing. I am not a doctor, I am an engineer. But the subject of the distribution of the money basically comes down to the people who are making the decisions. The people who are making the decisions basically live in the urban sector, and this is why the money is being spent that way. This is a vast generalization. I can demonstrate with figures that it is as cheap to build the rural water supplies as it is to build the urban water supplies.

DR. GREIM: Dr. Smeets, when you presented a resolution of the Commission on European Drinking Water Supply, you told us that there was a recommendation that artificial softening of drinking water is not recommended. I really wonder why this has been included, because in Germany, I know at least several places where this is already installed into the drinking water supply. This raises the whole question of what is the recommendation in the Common Market?

DR. SMEETS: Well, I did not say it in this way. To this particular point of the softening of the drinking water, that was the result of the colloquium I mentioned, we issued a draft on drinking water directives. What we did in the draft was to put minimum required concentrations of certain elements as one point you made. The second point you raised is that there are quite a number of softening apparatus already installed. This is true. But, what I said in my paper was that the installation of this softening apparatus should not be encouraged for the time being, since although we do not yet know the reason, there is an inverse relationship between the hardness of drinking water and mortality, particularly by cardiovascular diseases. But as long as we do not know the precise reason, of this inverse relationship, we should not encourage the use of softening apparatus.

DR. ERNST: A short remark, and an answer to chlorine. In the meeting on which I reported, there was a report on the Iraqian waters, and that is a case where they really solved immense microbiologic problems, just with chlorine. Of course, Iraq is rich enough to buy and use all of the necessary equipment for this sort of thing.

MR. JACKSON: I am Executive Director of the Asbestos Cement Pipe Producer's Association. I am not a medical doctor, I do not consider myself a scientist, I consider myself an engineer. I think there is one topic that we have been alluding to, and that has not been brought out - the whole idea about sampling at the treatment plant and then, after the distribution through piping materials. Asbestos has become a very controversial issue here in the USA. We firmly believe, and we are carrying out studies now that the contribution of asbestos, through asbestos cement pipe, and the contribution of any trace element, be it through lead pipe, cast iron, or plastic coated pipe, is a function of the aggressiveness of the water. There has been no attention paid to treating the water as it leaves the treatment plant, in terms of its aggressiveness to the distribution system, which is a direct function of what materials are going to get into the water supply, from the time it leaves the treatment plant where it meets the quality water standards and that which the consumer sees at the tap.

Now, we have developed guidelines, based on a modified index, which is simply the pH, the alkalinity and the hardness which seem to correlate very well with, in the case of asbestos cement pipe, any degradation of the inside of the pipe barrel. We believe that those aggressive conditions will also apply to other piping materials.

MR. CLARK: I just want to say a word about the Citizens' Drinking Water Coalition. You people are the very credible scientists in this business. You wish to reach you own decisions, but in the end, the public needs to be involved, and the Citizens' Drinking Water Coalition is an attempt to get many of these conclusions in frontof the public. We have an Executive Committee, with Robert Harris of the Environmental Defense Fund, as an advisor, Gus Speth of the Natural Resource Defense Council, and David Swick of the Clean Water Action Project. Carol Jolly of the National League of Women Voters, Education Fund, is on the Executive Committee. Lou Sirico, of the Public Interest Research Group and Joe Hyland of the Environmental Defense Fund.

We are attempting to have meetings in which we bring in representatives of the public. We have some 45 groups around the country that get our periodic newsletters. We are attempting to attend the National Drinking Water Advisory Council meetings and testify. We have made quite a study of the state laws that apply to drinking water. And we are trying to mobilize the citizen groups to be concerned with this issue, since in the end, the public must pay for what happens.

We thank Dr. Truhaut, particularly, for his emphasis that one must consider human beings as the most sensitive species for untested toxic chemicals. For indeed, we must not run involuntary experiments on humans. Hence, the repeated emphasis of those who accept continued pollution that there is no proof of harm to humans, is not an acceptable scientific argument, and we would appreciate the reinforcement by the International Academy of Environmental Safety that animal data must be used, when human data is not available, as a basis for judging the dangerous levels of toxic chemicals. We appreciate Dr. Truhaut's emphasis that to ensure the safety of humans, toxicologists have used a factor of safety of at least one hundred of the exposures that probably will not injure humans in comparison to the exposures that will produce a just detectable effect in the most sensitive of all tested animal species.

We do wonder if the factor of safety of 100 is large enough. Dr. Truhaut and others have emphasized that safe levels should be set by toxicologists and other established scientific workers, arguing in the technical literature. We do encourage you when you reach you conclusions, and as you reach your conclusions, to involve the public as much as possible. We are concerned that if you take the appearance of saying, "Tut, tut; leave it to us, we will tell you the final truth," and say it as briefly as that, the public will not be satisfied that you are totally open with us. We are particularly concerned with conflict of interest situations, since in general, the scientific establishment is supported by the business establishment, and so on. The funds flow from the source of money.

We are concerned with the benefit-cost analysis approach, since so frequently the comparison is a cost of things in comparison to a cost affecting people. The benefit, all too often, is benefit to some industrial group, and the cost is some deferred cost to people getting sick later on. I, myself, in my safety work am urging the use of what I call the humanitarian multiplier, a factor of ten for the costs involving people, in comparison to the costs of things.

DR. MRAK: If there are no further comments, we will adjourn this morning's session.

Lunch - Speaker: Dr. Emil Mrak

Dr. Buckley outlined some things for me, and some titles, and he gave "Some Thoughts on International Involvement." The EPA is interested in joining with other countries in solving problems. They are very intensively interested in this area. EPA is working with our neighbor, Canada, especially with respect to the Great Lakes. I knew a little about the mercury problem, but I did know they worked on it. In fact, I think it was a Canadian who first got us to look into the mercury problem in the Great Lakes.

EPA is looking for special opportunities to take advantage of and the harmonization of regulations. This morning you heard a little about different standards and different regulations. EPA is interested in doing what it can to harmonize these so they do not vary all over the lot.

Now some of the things they participate in. Inter-governmental organizations such as WHO and its many committees. Then there are environmental programs in FAO and UNIP, and, too, the Pan American Health Organization, PAHO, ECA, UNESCO and OECD. There are also activities with NATO, and so it goes. So there is a lot of international activity on the part of EPA. Sometime, I would like to see a conference where EPA could outline what it is doing on the international level. I do not feel many of us, including myself, really know.

They also have bilateral agreements with other countries, for example, Canada, and I believe, Japan. They have an agreement with the Federal Republic of Germany and the USSR, and in fact, the Administrator of EPA, Mr. Train, is Chairman of the environmental agreement there. There are foreign programs with Poland, Yugoslavia, United Arab Republic, Tunisia, Pakistan and India, involving funds and individual technical assistance.

When I retired as Chancellor, I had many phone calls and offers, and one of them was to go with Vice President Rockefeller on his missions to Latin America, and this I accepted. About two days later, I had a call from the Secretary of HEW and he said, "We are in pesticide trouble, will you chair a commission?" And I said, "I want to, but I am going with Mr. Rockefeller - but, I have an idea." You heard Dr. Lindsay yesterday. He was then my assistant chancellor, later Associate Commissioner for Science for the Food and Drug Administration. I said, "Why don't we let him handle this until I get back so I can become involved in both of these activities." And that is what happened. Dr. Lindsay did an excellent job on the Commission on Pesticides and the Relation to Environmental Health and the Environmental Aspects of Pesticides. A 700 page volume came out of this study. Although several thousand copies were printed, it is now out of print. Out of this report came some interesting results. In the report, there were sections on carcinogenesis, teratogenesis, and mutagenesis. The importance of teratogenesis and mutagenesis was really emphasized in this report.

We also considered general and public health aspects, needs, uses and non-target organisms.

After the report came out, Secretary Finch went about implementing the recommendations. One of them was that he set up, for at least a short time, an advisory committee on pesticides. The Secretary's Pesticide Advisory Committee, called "SPAC" reported directly to the Secretary. But about that time there was a movement to create a new agency - the Environmental Protection Agency.

When it was decided to form an environmental protection agency by cannibalizing other agencies, Dr. Talley was on the working group to look at this. Dr. Talley, now the Assistant Administrator for Research and Development, was involved with EPA from the very beginning. It is interesting to me that a critical mass was needed insofar as funds were concerned. Since the EPA was created, it was necessary to do enough cannibalizing to reach the critical mass of several billion dollars.

The other thing I found interesting, was the minute that various government agencies found that they were going to lose staff and support, they started shuffling personnel so as not to lose its best personnel. And so, when agencies are created in this manner, it may take years to get over such effects.

Eventually, the Secretary's Pesticide Advisory Committee was moved over to EPA, and the name was changed to the Hazardous Materials Advisory Committee. Dr. Allen is Executive Secretary of the committee. Of the people here, Dr. Golberg, Dr. Lindsay and I served as members of the committee.

This committee became very active in studying hazardous materials and became known as the Hazardous Materials Advisory Committee. It was a very active committee. The EPA inherited about ten committees involving about a hundred people. These committees were concerned with air, water and radiation.

About that time, Dr. Greenfield, who preceded Dr. Talley as Assistant Administrator for the Office of Research and Development, decided they should have a science advisory board. Our government moves slowly. This was about three years ago. First, it must be approved internally, and then by the Office of Management and Budget. Only last January did it finally get approved within EPA.

It was in the mill for over a year, and then finally approved by the Office of Management and Budget. It has absorbed all the other committees and then created an Executive Committee.

It was not possible to have a hundred or more people representing all the old committees meeting with the Administrator, so an Executive Committee, consisting of the Chairmen of Subcommittees, plus a few people at large, was formed. When some one asked me, last January or February, who was on this Executive Committee, and what was it I said, "It's me." Because at that time, I was the only person on the committee. How I got there, I do not know, but I did get a letter from the Administrator. So, the Science Advisory Board was created and the various advisory committees came in under it.

Eventually, when the Science Advisory Board was formed, new committees to fill the needs of the agency were created. The first one created was concerned with ecology. Dr. Ruth Patrick chairs this committee. She is a limnologist, a very talented one and very capable. As I told you earlier, she could not be here because she had an emergency operation.

This committee has had several meetings. It is reviewing the programs of the some 39 laboratories and facilities scattered through the country that are carrying on research and monitoring for EPA.

The committee is also working with other committees to review reports such as the one on the disposal and utilization of sludge and sewage waste. Sludge is a real problem. We have improved our sewage, and in doing so, created sludge. Now, what do you do with the sludge? It cannot be dumped into the ocean and some of the states will not permit it to be dumped on the land. You have heard at this meeting about cadmium, and there is cadmium and other minerals in sludge, so it should not be put on the soil. The question is, what to do with it? That is one of the problems that is receiving a lot of consideration now.

Another committee recently created is concerned with environmental health. It is chaired by Dr. Norton Nelson. To me, it is tragic that it has taken so long to create this committee concerned with health, for strange as it may seem, EPA has concerned itself with environmental health matters from its beginning. For example, the use of DDT was for all intents and purposes banned on the basis of potential health harm, but there was relatively little scientific input to support such action. Strong health advice is needed and with this new committee EPA will get it.

Another new committee that has been created under the Science Advisory Board is one on environmental measurements. You have heard a great deal about the measurement of different things. You heard Dr. Egan talk about how accurate and inaccurate we can be. Should we standardize the people who make the measurements, or should we standardize the measurements, or both? So, we now have a committee concerned with measurements and monitoring, and it is chaired by Dr. Rossini of Rice University.

Another committee is concerned with the movement of transformation of environmental pollutants. A good example, and one of concern to this group, is acid rain. You have heard a little about it, and it is important. In New England, people monitoring the lakes and water there now find that the pH is going down. I do not know about monitoring for the minerals but I was interested to hear at this conference that lead and other minerals can come out of the air into water. This is probably something that will be considered. At first, those monitoring for acid rain thought only in terms of locating and controlling power plant emissions within ten miles of a water supply. Now, however, they are thinking in terms of two hundred miles or more.

Another committee will consider environmental pollutant movement and transformation and this is chaired by Dr. Montrol of the University of Rochester. How are we going to control such things? How are we going to clean up smokestack emission or even automobile emission?

Now, about a few of the activities that some of these committees have had. One of the studies was made on hexachlorobenzene, a rather interesting chemical. A representative of one of the chemical companies came to me, almost at the birth of the Environmental Protection Agency, and said, "Look, we have observed that in burying discarded waste products from the manufacture of chlorinated hydrocarbons, HCB and hexachlorobutadyene may be leached, contaminate streams and possibly ground water." This company worked out a new system of disposal and felt EPA should know about the problem and what the company was doing about it. The company desired to work with EPA, but EPA was growing and trying to congeal at that time, so nothing much was done until the US Department of Agriculture, analyzing the fat in the tail of a cow, found HCB. Where did it come from? It came from the buried waste near pasture land in Louisiana. It was also found in water so a study was made of the situation and the contaminant finally eliminated.

A study was made of the emergency use of DDT on cotton and another on nitrogen in the environment. Of course, a study on the assessment of the health risks of organics in water was made by Dr. Murphy and his committee.

Currently, a sub-committee is reviewing a document on the environmental aspects of asbestos. Dr. V.K. Rowe heads this committee.

The Science Advisory Board has also established ad hoc committees to make studies in certain areas, one of which is concerned with sulfates. Another was to study the EPA's program on the Strategic Environmental Assessment Program, known as SEAS.

The Executive Committee of the Science Advisory Board meets about four times a year and at these meetings, considers reports of such committees. The Administrator usually attends these meetings, and this, to me, is very important. Mr. Train spent a lot of time with the committee. It is a new and important development. He is indeed interested in the scientific aspects of his many problems. Talley, of course, has attended right along. Naturally, he is interested in the scientific comments and advice of the committee.

Now, I might comment on one or two other things that are going on. Life for the EPA is not easy, as you can well imagine. Congress is interested in what EPA is doing and this consumes an enormous amoung of time and effort. Congress, at times, has questioned the decision-making process and accordingly, appropriated and put into the EPA budget, $5 million earmarked for the National Academy of Sciences, National Research Council, to study the decision-making process in EPA. The study is going on very intensively now. Mr. Train talked with some of those making the study, and I am certain he appreciates their views and advice.

MR. CLARK: Some of the public interest groups are concerned with the decision that apparently has been developed within the EPA, that the panels of the Scientific Advisory Board, because they suggest alternatives but do not deliver advice to your Science Advisory Board, do not have to comply with the Federal Advisory Committee Act. And, we are concerned about that.

We see some of the decisions really developing on the basis of scientific information being collected by your panels. We do urge you, wherever possible, to treat the panels with the same openness that the Advisory Committees now are treated.

DR. MRAK: I have heard that discussed, and I will be perfectly honest with you. I dodge it - because I think they have enough attorneys in EPA to handle these things. The operation of my group depends on the advice of the attorneys at EPA. They are well aware of this. Now, if there have been things going on that shouldn't be, I guess they ought to know what to do about them.

DISCUSSION NOTE: At this point, many questions were asked of Dr. Mrak concerning the function and membership of the EPA Science Advisory Board. Since this was not in order for this Forum, it has been deleted from these proceedings. The tape recordings are available.

CHAIRMAN'S SUMMARY OF FORUM

DR. COULSTON: Needless to say, we have had much discussion, some of it very stimulating, and some of it a little mundane. But, in general, it has been a rather provocative and exciting meeting. We come now, this afternoon, to yet another phase of the problem of drinking water quality. We have alluded, during the past two days, time and time again, to the mysterious presence of nanogram amounts of substances that all of a sudden have been discovered in water.

I think chemists have known for hundreds of years that water is a pretty good solvent, next to alcohol, of course. And when you mix water and alcohol together, it is perhaps the best solvent system of all. Therefore, why shouldn't there be things in water? Water can dissolve almost anything in minute amounts. But all of a sudden, we have a big cry that we are discovering chemicals in water.

Yet, as part of our process and part of our socio-economic system in our civilization, it behooves us to satisfy the public that he is not at risk, and that someone, somewhere is protecting him from some kind of an unseen hazard. The consumer has this right, and consequently, governmental agencies and other groups respond to this. Unfortunately, the problem usually crystallizes to the right or to the left, sometimes just like politics, and may even become political.

What we are seeking are facts. The facts as we know them, and as I said the other day, unfortunately the scientists do not know all the answers. When a "scare" or a calamity occurs such as the question of nitrosamines in the air or water, then the press and the news media in general, react. The have a right to. Because they have been told that they are consuming a carcinogen. Then, somewhere along the line, the regulatory agencies are going to react, and go far beyond our scientific ability to answer the questions. And then, comes the most horrible result, in this sense: useful products are banned or restricted and the public is confused.

If the most sensitive species can produce a tumor, which during a lifespan study can become a cancer in a few animals, at a fantastically high dose, why take the risk? Ban it.

Is it possible that the chemical can cause cancer in man? The good scientists would say that they really don't know. The people to the left say, emphatically, "I know this is an imminent carcinogenic hazard and must be banned, immediately."

But, the fact remains, We, in this country, are spending millions of dollars trying to get answers to questions that science doesn't know how to answer. Is the chemical a carcinogen or not? What is a cancer? What causes cancer? And then, the concept creeps in that only <u>no</u> risk is acceptable.

Nonsense! To live in this world is to be always at risk. Since we cannot always make a decision based on animal data, we must begin using epidemiologic data from human studies.

We are asking almost, what is life? We are getting into a ridiculous situation where we will impede the growth of our civilization, if we push this quest for answers that cannot be given.

The economics involved for a no-risk environment are so great that we could easily bankrupt the nation. For example, why shouldn't every water treatment plant in this country use charcoal filtration? You would reduce the contaminants in water to a point where the major problems would disappear, but the cost of doing this would be billions of dollars a year.

As has been mentioned, another alternative is to have two kinds of water, one for drinking and one for industrial and washing and the kinds of things you do around the house.

I don't see anything wrong with that, except again, you are up against a soci-economic equation. In my house, I have an old-fashioned well, 100 feet down. It produces nice hard water, 40 parts hardness, and when you cook with it, it coats the pots and the tea kettle. But we drink this water in preference to the city water, which we then use for all our other purposes. Obviously, the dual water system has great merit.

What I have just been discussing is in reality an introduction to the final part of this Forum. Are the chemicals that we find in the drinking water and in the water supplies a hazard? Can we establish a benefit-risk relationship to these chemicals? Should they be banned, restricted or removed from water?

Obviously, anything that is dangerous should be taken out or reduced to an amount that is harmless to man. In the final analysis, a judgment must be made based on some kind of scientific evaluation, which must then be judged by the politicians and lawyers, and finally, by the public. They will judge what risk they will take and what is acceptable to them in terms of benefits and risk.

The individual wants to know what is good or bad for him. Do you want to put fluoride in the water supply for your children, or do you want to put a drop or two in the orange juice in the morning, and feed it to your children? If it is done this way, it means you don't have to get the fluoride. A perfectly sensible way to do it. Put it in the toothpaste, if you wish. Then, you can buy that toothpaste or not, depending on whether you want the fluoride or not.

These are personal judgments and we must try to protect our rights to this judgment, or else we will be regulated literally to death.

Now, Dr. Leon Golberg will speak first this afternoon.

WATER QUALITY IMPLICATIONS OF HMAC (EPA) REPORTS ON AGRICULTURAL CHEMICALS

Leon Golberg

I feel sure that the Safe Drinking Water Act of 1974, and the question of national interim primary drinking regulations, have been adequately discussed, with reference to standards, specifying maximum levels of drinking water contaminants, and monitoring requirements for public water supply systems.

My task today is, in part, to prepare the ground for Dr. Murphy's presentation of the evaluation of health effects of chemicals in water. I shall do so, by reviewing the work of the Hazardous Materials Advisory Committee (HMAC), and the ground for this has been prepared for me by Dr. Mrak, who is the best authority to speak on this subject.

Of him, it has been truly said that he seldom is wrong, and never in doubt. And, we had experience of that during his supreme leadership of the HMAC.

At one time or another, HMAC has concerned itself with virtually every effluent or other water pollution problem. And this brief review is intended to highlight some of the committee's contributions in this area, prior to the studies to be reported on by Dr. Murphy.

To summarize the functions of the committee, very briefly: what emerges is that the committee has reviewed the research on monitoring activities within EPA; has helped to identify emerging environmental problems, with regard to hazardous materials; and, has considered how such problems might be contained or brought under control.

Agricultural and related chemicals have always constituted an appreciable proportion of the committee's concerns. Apart from the importance of this subject, to the public and the environment, the committee's special interest in pesticides may be traced back, as Dr. Mrak indicated, directly to the origins of the committee, right back to the Secretary of HEW's Commission on Pesticides and their relationship to environmental health. So, I will try not to go over that part again; but, for the benefit (particularly of our guests from abroad), I should like to quote an explanation taken from a review last year, "To analyze the power and function of the Hazardous Materials Advisory Committee of EPA, it is necessary first, to describe the role of government advisory committees in general."

"The central problem in appreciating the power of advisory committees, is that, although advisors speak with authority, as respected experts, they do not have authority to command or act. There are no concrete monuments to the decisions or acts of advisory committees. They do not, by themselves, cause pesticides to be registered or discover chemicals which may be used in place of more harmful ones."

"The power of advisory committees is exercised obliquely. The affects of their power are discernible only in the decisions of those they advise, and by comparing those decisions to that advice, as expressed in the studies, reports, letters, minutes of meetings, and comments, which, without legal affect themselves, are the traces of the advisory process."

"The advisory committee shall not command. It must consider and convince. Responsibility for decision-making and action, remains with the agency officials."

I would be less than frank with you if I did not make it clear that many of EPA's most far-reaching decisions on what were considered to be hazardous substances in the environment were made without reference to the HMAC, or in direct contrdiction to the committee's opinions and advice. Such are the realities of political and regulatory life, and we have to recognize them.

None the less, it is useful to recall, for this conference, some of the more notable experiences in the course of the committee's activities, as they relate to water quality and some of the other issues discussed here.

the question of pesticides in the aquatic environment is one that was addressed by the committee. The subject of pesticides and pesticide container disposal has been a constant preoccupation of the committee since the beginning. Accordingly, HMAC was asked to review and comment upon two EPA reports, prepared by the Office of Water Quality Programs, in fulfillment of the requirements of the Water Quality Improvement Act of 1970. Section 5.L.2 of the Act, directed EPA to bring together for the purpose of adopting standards, the scientific knowledge necessary to develop water quality criteria for pesticides. Under this directive, EPA was involved in increased research on the effects of pesticides, on the search for less harmful pesticides, expanded monitoring and investigation to identify critical areas.

The same section authorized the pesticide control study, which was described as a study and investigation of methods to control the release of pesticides into the environment. The study had to include examination of the persistency of pesticides in the water environment, and alternatives thereto.

While these document, prepared under this regulation, did not represent a quantum jump in scientific knowledge, they did serve a most valuable function by assembling for the first time, and in an exhaustive manner, the widely scattered infomation on diverse areas of concern in the field of pesticides in water.

Over and above the critical review and analysis of EPA reports of this kind, HMAC was asked to undertake detailed examinations of specific topics in the area of agricultural chemicals. Some have been mentioned by Dr. Mrak - I shall briefly review three of the committee's reports, dealing with toxaphene, with herbicides, and nitrogenic compounds in the environment.

As we approach toxaphene, in the course of the survey of the status of this material undertaken by the committee in 1971, attention is drawn to the variations in the rates of dissipation of this remarkable mixture of chlorinated camphenes, comprising in all, about 150 different compounds. In lakes and ponds, the material was said to be capable of persisting for up to nine years. Other data yielded a half-life in water that could be as long as six years. It was difficult to reconcile these assertions with the fact that toxaphene toxicity, biologically, does become sufficiently reduced to permit lakes and ponds to become habitable for fish after toxaphene treatment. The time required depends on the usual diverse factors of temperature, rainfall, surface run-off, microbial populations, exposure to UV light, etc., but certainly, is very much less than the six or nine years that have just been mentioned.

Difficulties and uncertainties associated with the quantitative analysis of toxaphene residues, probably render the early data of questionable validity. We do know, however, that in waters treated with toxaphene, aquatic plants, invertebrates and fish, all accumulate the material. Some fish may contain several ppm of toxaphene in their tissues for as long as a year after the treated waters are no longer toxic to them. Thus, biomagnification of toxaphene residues does occur, but apparently to a lesser degree than with most other organochlorine pesticides.

The HMAC report stressed the urgent need for more information on biomagnification and persistance under conditions of normal usage of toxaphene, with special reference to the possible build-up in food chains, under aquatic conditions. It was pointed out also that information was lacking on the degradation by photochemistry of toxaphene in the soil and water, and under biotransformation, in animal tissues.

The report was timely and clearly pointed to a need for more data in the face of increasing use of toxaphene. It is interesting to review some of the aspects following implementation of the committee's recommendations. The new evidence does not suggest a build-up of toxaphene in the environment. On the contrary, monitoring studies show no accumulation in the soil, in areas of heavy use of toxaphene, and little or no transfer from such areas to non-crop sites. The reason probably stems from the fact that toxaphene is not used as a soil insecticide, but rather as a foliant spray with rapid dissipation. Surveys of stream and surface waters in the USA have not revealed the presence of toxaphene by analytical methods, whose lower detection limit was 1 ppb.

The EPA monitoring study of New Orleans water supply did not detect toxaphene, despite the intensity of use of this material in the drainage area of the states immediately above New Orleans. Other evidence that agricultural use of toxaphene has not resulted in water-borne dispersion of the chemical, was provided by the mollusk monitoring program carried out during the years between 1965 and 1972.

Detoxification of toxaphene is characterized by irreversible absorption of toxaphene on soils and other particulate matter, as been previously observed in fresh water lakes.

In addition, chemical degradation of toxaphene in sediments occurs probably under anaerobic conditions. So, you have the situation where photodecomposition of toxaphene, absorbed dredge soil, may involve an interaction, because thin films of toxaphene exposed on glass in the laboratory do not shown photodegradation. These are some of the complexities that one has to be aware of, in trying to extrapolate laboratory data to what actually happens in the field.

Consideration was given to herbicides, including the ones that are used to control aquatic weeds, to the greatest extent, in this country. As far as aquatic plant control was concerned, the report concluded that in all but a few instances, the efficacy of herbicides far exceeded the effectiveness of presently available alternatives. Nevertheless, the control of aquatic plants is difficult to manage, and it is even more difficult to assess the ecological impact of control.

The major environmental problems arising from the use of herbicides involve the operational techniques which have been developed in recent years and have progressively reduced the movement of herbicides by drift. Persistence in soil is not necessarily undesirable. In fact, this property may be a considerable asset. Such persistence need not affect adjacent bodies of water. Pollution of water was considered in terms of both the individual parent herbicidal compound and it's metabolites, and products of degradation, including photodegradation.

A number of herbicides that are persistent in soil are less persistent in water. For example, picloram is subject to decomposition by ultraviolet light in surface water with a half-life ranging from 2 to 41 days.

In the context of the particular herbicides considered, health effects were not found to present problems of any magnitude, either for applicators or as regards the ingestion of herbicide residues. Undoubtedly, however, there is a need for further studies of long-term effects in several instances.

I turn, now, to a report on nitrogenous compounds in the environment, which the committee undertook early in 1972. This comprised the study of the sources and effects of nitrates, nitrites, ammonia and other nitrogenous compounds in the environment. The report was issued in 1973. In reviews pertaining to sources of nitrogenous compounds and methods for their control, the following categories were covered: municipal discharges into water, as solid waste; discharges into the atmosphere; discharges into the environment from crop production; discharges into the environment from animal wastes; and major industrial discharges.

In regard to nitrogen-containing fertilizers, the report stressed that although the effect of fertilizer use or misuse on nitrogen levels in waterways had been studied extensively, there was still a need to define more clearly the nature and causation of a regional pattern of nitrogen concentration in water.

This was especially true of ground water, which had received little attention. The report laid considerable emphasis on the detrimental effects of nitrogenous compounds in the aquatic systems, particularly to many forms of aquatic life now occurring in waterways. In addition to the nitrogen contributed by livestock and human wastes, industrial discharges provide nitrogen compounds, possessing special significance.

One has to bear in mind the potential for interaction of the nitrate/nitrite in the environment, with compounds of industrial effluent to form nitrosamines and nitrosamides, which in many cases are potent direct or transpercental carcinogens in animals and probably also in man. Several features of this interaction between nitrite and organic compounds need to be taken into account today, and I hope I am not repeating what has been said on previous days. But, I think it is very important to stress these points.

Another topic I would like to refer to is the hexachlorobenzene incident, which has already been mentioned by Dr. Mrak. I think the interesting feature of that was that the industrial representatives who came before the committee stressed the fact that the chlorinated residues from stills or from other solvent manufacture processes that were buried in the ground need not comprise hexachlorobenzene of hexachlorobutadiene.

There could be all kinds of chlorinated residues, but that in time, they would be converted into this most thermodynamically stable form. And, it was very interesting that in the New Orleans incident that Dr. Mrak referred to, where the residues were found in cattle, this was precisely what had been happening. They had been burying the residues, but what was actually monitored and found was hexachlorobenzene. And, in actual fact, although the water itself contained very little HCB, the silt at the bottom of rivers did contain a substantial amount. For example, stream sediment samples in that area ranged from 53 to 706 ppm of HCB, and crayfish slightly outside the area contained .16 ppm HCB in fat.

Another point to be mentioned - we should have learned a lesson from Louisiana about the hazards of HCB distribution. But it is still around. Dr. Yang, in our own laboratory, has recently found that animal diets can contain substantial amounts of HCB in a most erratic manner. For instance, Purina Monkey Chow has been found to contain up to 21 ppm of HCB. And it seems that the source of this material is fat, probably some kind of waste fat that is incorporated in the animal diet, so that, within any one batch, you will find quite erratic figures in different parts of the same sack. This might not sound very important, seeing as it is being fed to animals, but I would have you consider, for one moment, the implications of a situation like this when people are studying the toxicity of PCB or other materials, which are a source of great concern from an environmental point of view. If the basic diet already contains 20 ppm of HCB, you can imagine what kind of artifactual results can be obtained in the animals. One has to really be very much more conscious than any of us have been in the experimental field, of the need for very thorough and consistent and repeated analysis of the animal diets, or otherwise, there has to be some standard of quality control set up from a central point to insure that these artifacts will not arise.

Finally, I would like to say a few words on the subject of interactions, and this again, was something that the HMAC has been very much interested in. The multiplicity of natural and synthetic contaminants in water raises the question of toxicological environmental and other forms of interactions between these compounds. The interactions may occur between different pesticides applied to the same area. In fact, to the same top layer of soil, and then washed into rivers.

Or, the interacting chemicals may also include those from natural sources, as well as pecticides in the products of chlorination of both groups. That is, chlorination of water containing both pesticide residues and naturally occurring compounds.

The modifying affects of natural factors in the environment also play a key role in the interactions. The 1972 National Academy of Science study on research needs and water quality criteria, pointed out that the role of organic materials in correlating, stabilizing, and transporting toxic heavy metals in the aquatic environment is virtually unexplored, as are the possible alterations in physiological responses. The study stressed that the greatest need is for research on the interactions of suspended matter with toxic substances. And that, in addition to studies on the biological effects of toxicants, research is urgently needed on the interaction of toxicants with other compounds in the environment, derived both from living and non-living sources. Of particular concern is the response of aquatic life to exposure of culminations of toxicants where synergistic or antagonistic effects are possible.

HMAC has, at one time or another, interested itself in a variety of interactions of this sort. Obviously, environmental science is only at the beginning of the long road that will lead to adequate understanding and effective action to minimize any hazards that may be present in water, particularly those stemming from the use of agricultural chemicals.

DISCUSSION

DR. COULSTON: Dr. Golberg's remarks just prove how complicated things can be, and are. In this light, I would like to break policy, because of the importance of some of the things that have been said. I would like to allow a brief discussion at this point, and I am sure some of the chemists have statements to make.

DR. EGAN: Nitrosamines have been mentioned a number of times, and I have stressed earlier, the great importance of positive identification of not only trace nitrosamines, but other trace toxic substances.

Perhaps I need only say that, by way of illustration, it is not sufficient merely to accept one single chromatographic tracing as evidence, as positive conclusive evidence of a particular nitrosamine, especially when that trace consists of very many peaks, only one of which you think you are going to attribute to the nitrosamine. It is a very simple matter, but I think perhaps it is worthy of mention at this point.

DR. GOLBERG: I think that there has been ample experience in recent years to support what Dr. Egan has been saying. And, I think although we ourselves are not in the area of measuring nitrosamines, I think every one is aware of these snags, and of the need for the utmost care to get supporting evidence by mass spectrometry and other means.

DR. JUNK: I was quite interested, and I may have misinterpreted what you said, Dr. Golberg, but a good share of our water supply, particularly through the Central States, the Mississippi River, also true of many other rivers, have relatively high concentrations of atrazine. It is very widely used. And, did I read you correctly, that there was experimental evidence to indicate that there was a reaction between nitrate levels, which are also quite high in the same water supplies, and atrazine, leading to nitrosamine formation?

DR. GOLBERG: Yes, both with atrazine and simazine.

DR. JUNK: I might comment that that is very interesting, because I do know some people who are trying to simulate that possible reaction in the laboratory, with many different native water supplies. So far, they have not been able to demonstrate in the laboratory that nitrosamines can be formed from atrazine and nitrate. I am glad to hear that.

DR. EGAN: I do not know the answer to the question, but it

was, did you say, Dr. Golberg, that nitroso compounds were formed, or that nitrosamines were formed?

DR. GOLBERG: I said nitroso compounds. And, of course, one has to distinguish between a system, an experimental system, where you are deliberately seeing whether the reaction will go and an investigation of whether the nitrate in water will do it.

DR. SOMERS: I just wondered about the pH, Dr. Golberg, of those systems you were talking about. I was suprised it was low enough to form nitro compounds.

DR. GOLBERG: These again, are under experimental conditions. I have the reprint on the work, and as far as I remember, it was pH 5.

DR. COULSTON: We have had quite a discussion on whether nitrosamines could really form in water or not. It has been pointed out that the pH is very critical or else the reaction cannot go, except below pH 3. And so, I think that is what they are pointing out. But it would be difficult to imagine these kinds of pH in a river or a living body of water. Perhaps in an animal stomach, the pH can be low enough.

DR. GOLBERG: I would merely say that I think, as evidence develops, we are beginning to see that there are situations where this critical pH can change, because of the presence of other substances. And, I was merely indicating that we have to keep an open mind on this. That situations may arise where there are other substances present simultaneously, which make possible the nitrosation, or inhibit nitrosation.

DR. SUESS: I would just like to support a point made. I am not aware of, that is, I have not heard about formation of nitroso compounds in water. But, in stomach, which is really the problem, since we do not know what the case is.

DR. COULSTON: Let us not discuss the problem of formation of nitrosamines in the stomach. We are only talking about this substance in water. Can they form in water? As you know, just the other day they reported from the EPA.

DR. ERNST: Dr. Golberg, you reported on photodegradation by u.v. I was astonished that this occurs in surface water, since the u.v. usually can only go a few millimeters into the water. Is this really an important part of the whole turnover?

DR. GOLBERG: Of course, I cannot claim to be an expert on this, but it seems to be a well established fact that it does have a big influence. Now, perhaps in clear water, it does penetrate deeper than we think.

DR. KORTE: It was shown, during the last year, that toxaphene and all similar chemicals, are degraded not so much by photodegradation reaction rates in water, but in the air. In absorbed phase in particular, this has been shown with daylight, diffused daylight, completely.

In 1969, on the occasion of the First International Symposium on Environmental Quality, the meeting agreed to some statements, when to use chemicals and what are the purities. And, everyone agreed the first point should be that one should definitely know the structure. One also should know the purity of the chemical used. This was, actually, the time when DDT was banned first in Sweden and the Swedish representative reported this. Since then, we did some experimental work with toxaphene, and we identified positively 14 percent of the isomers chemically, and we have some doubts that any one has a realistic and good method to find residues in environmental samples.

I cannot see a possibility of how to measure a chemical when I don't know the structure. I would not argue for or against toxaphene. But, if we discuss scientific bases, then we also should make clear that, I don't find it scientifically correct, to replace the well known DDT by toxaphene. And, if we look at the production figures on a yearly basis, over 100,000 tons of the chemical is used a year.

DR. COULSTON: You make a very good point. In other words, if you don't know the structure, how do you know what the metabolites or even the end products are? The reason you don't know, is that toxaphene is a mixture of many chemical molecules. In fact, the standards for toxaphene vary greatly and they may be composed of many individual compounds.

DR. EGAN: At least 150 different compounds.

DR. COULSTON: The point is that in the USA, with DDT restricted or gone, and with dieldrin and aldrin banned, and with heptachlor and chlordane about to be banned, we only have left toxaphene. And, following this line, toxaphene may also go. There is very little left. There is no substitute any longer for use in good agricultural practice.

DR. JUNK: I would like to reinforce the comments made by Dr. Coulston. It is rather interesting that we choose to ban a chemical once we accumulate the ability to detect that chemical. Once we accumulate the ability to determine the possible adverse effects of that chemical, it is banned, and maybe I would question whether there are substitutes for the chlorinated hydrocarbons.

I am certain that the pesticide people and herbicide people would say there are substitutes. But, what comes on the market, then, is a whole host of substitutes, of which we know nothing. And, it will take the analytical chemist 10 to 15 years before we figure out how to analyze for them. It will take a toxicologist another 10 or 15 years before they determine the health effects.

And the whole situation leads to regulation, where the end result, after a period of time, may lead to a worse situation than what we had before the regulation was imposed.

DR. EGAN: What you have said is very important. However, I was not aware that DDT had been banned by society, yet.

DR. COULSTON: This is an international conference, and regardless of what is said for USA, the fact is that DDT is widely used today worldwide, because it has to be used in certain parts of the world.

DR. JUNK: My only point is that scientists should never propose to replace a well known chemical by a structurally unknown one.

DR. COULSTON: I agree with you. Let us have great caution, before we remove something where we have many years of experience, even as to human health, let alone the environment, with something that is relatively unknown.

This was one of the problems of NTA. I, myself, don't want NTA back, but it might surprise you, Mr. Clark, why I don't want it back. I don't want it back in the United States because it entails the transportation of thousands of gallons of cyanide on our transporation systems. I have steadfastly argued against NTA for this reason, not that it is a hazard to the environment. But think of what I have said. I would rather ban it on the grounds that the ingredients are too dangerous and that I don't want special trains or tank cars going over our highways at 30 miles an hour with the possibility of a terrible accident occurring. This is my reason for not promoting NTA, and I think that may surprise some of you.

DR. GOLBERG: I want to respond very briefly to two points. First of all, as regards what Dr. Korte mentioned, I agree entirely. I always feel very uncomfortable, trying to assess the safety of a material that is a mixture of totally unknown composition.

But, I would remind you that there are many such materials, not only food additives, but in fact, the most widely used food color, caramel, falls exactly into that category. The only attempt that has been made to make caramel safer, in recent years, was the discovery that it could contain a highly potent convulsant agent. Now, we have laid down specifications to reduce this chemical in caramel. What the other hundreds of compounds present are, we do not seem to worry about.

The second point concerns toxaphene. I am not for or against it, but I would draw your attention to the fact that when the Mrak Commission was preparing its report, it had the Bionetic Study before it and in that study, a closely related material called strobane, had been found to be just as carcinogenic as DDT, and all the other chlorinated compounds.

And, curiously enough, no action has been taken by the authorities against strobane or toxaphene, in spite of the fact that the hazard to man, based on the mouse work, is presumably just the same as all the other things that they have banned.

DR. COULSTON: That is a good point. I think the caramel story is a particularly good one. It stresses a point that has come up time after time after time. Let us identify what is really harmful based on good experimentation, good science, and then take it out.

Let us move on to the next speaker, Dr. Sheldon Murphy from Harvard University.

PROBLEMS OF EVALUATING HEALTH EFFECTS OF CHEMICALS IN WATER

Sheldon Murphy

Dr. Coulston indicated that my primary area of interest is in biochemical toxicology. From the standpoint of my experience in evaluating health effects of chemical contaminants of drinking water, I speak primarily from the experience of chairing the ad hoc study group to consider the health risk of organics in drinking water that Dr. Mrak and others have referred to.

This study group was appointed and named last spring by Dr. Mrak as Chairman of the Hazardous Materials Advisory Committee. There were a number of individuals on that committee, expert in a variety of fields related to the problem. None of them, however, specifically were oriented towards drinking water problems.

But, what we had here was a unique and new experience of considering a broad problem from the standpoints of our respective disciplines, and I would like to acknowledge the fact that what I have to say today comes largely from the deliberations, the knowledge and the hard work that the members of that ad hoc committee devoted to the problem for a couple of intensive months on a part-time basis.

The members of the study group were: Dr. Harris Heartsler of the DuPont Company, an organic chemist; Dr. David Hole of the National Institute of the Environmental Sciences, a biostatistician; Dr. George Hutchison, Professor of Epidemiology at the Harvard School of Public Health, a physician and epidemiologist; Mr. Gregor Junk, whom you heard earlier in this conference, an anlytical chemist from Iowa State University; Dr. Benjamin VanDuren, a biochemist who spent many years in the area of carcinogenesis, from New York University Medical Center; and Dr. Gerald Wogan, a toxicologist who has been primarily involved in studies on carcinogenic materials, from the Massachusetts Institute of Technology.

These were the major participating members of the study group. We had excellent assistance from Dr. J.D. Allen and Dr. Tom Bath of the EPA staff.

So, although the basis of what I have to say comes from the work of many others, I do not think they should be held responsible for the remarks I make today, or the errors in interpreting their thoughts.

Now, I would like to review for you some aspects of the charge that this study group received on March 12, 1975. The agency had sought advice as to the significance of contaminants in drinking water, relative to potential carcinogenicity or other affects on humans, resulting from chronic exposure to these compounds. Carcinogenicity seemed to be the toxic effect of major concern. I might add that the basis of the request for assistance or advice, and I understand it, was partly related to the requirements of the Safe Drinking Water Act of 1974, which required to Administrator to study certain aspects within a relatively short period of time.

Secondly, because various public citizen groups and scientists had indicated concern about the potential, or possible association of cancer incidence with contamination of drinking water, largely as a result of an epidemiologic survey conducted in New Orleans.

The four compounds that the group began to study were benzene, carbon tetrachloride, bis-2-chloryl ethylether and chloroform. At the time the charge was given, there was a limited amount of data with respect to the concentrations of these compounds in drinking water, hence some approximate levels were suggested.

Of course, an estimation of risk considers not only the possible capability of a compound to produce toxic effects, but it also considers the possibility of exposure to concentrations that are capable of producing those effects. It is, therefore, important to know the kinds of concentrations we are dealing with.

Three other compounds that were identified for which information was requested, or for which consideration was requested, were beta-chloryl ethyl methylether, oxidecane, and thallic anhydryde. These, for one reason or another, had been suggested or suspected as possibly contributing to cancer, either directly or as co-carcinogenic material. In addition, the study group was asked to review a list of well over 100 organic chemicals that had been identified in drinking water in the USA - identified by various surveys of the Environmental Protection Agency, and others. The first list that we saw in March gave 162 compounds, and this has been steadily growing.

As you can imagine, we were asked to give a statement about the relative risks, and in conversation with agency staff, it was impressed upon our study group that we should attempt to give some quantitative estimate of risk. They would have liked to have had something that even though it would be an estimate, would give some quantitative relationship. We took seriously the charge to attempt to make a quantitative estimate, and as you will see later, we had to base such an estimate on very qualitative information. We did not have the facts we would liked to have had. We had a few, but beyond these it consisted of the application of certain assumptions and certain judgments.

The due date of May 1, 1975, for the report was very important, for that gave the committee approximately two months to begin to learn something about the problem, and to meet, consider it, and come up with a report. Obviously, with the large list of 162 chemicals, in addition to those that had been primarily identified, this appeared to be an overwhelming task, and it was difficult because most of the chemicals had merely been identified qualitatively, and actual concentrations were not known. I am not going to go through the list of the 162 chemicals, but I can tell you that they consisted of essentially every major classification of organics.

We were specifically asked, however, not to include in our considerations, pesticides, asbestos, and inorganic materials, because these were being considered by other groups. So, our considerations dealt with other kinds of organic chemicals that occur in water, including some groups other than the four primary and the three secondary, mentioned earlier. When we went over the list, we decided the chemicals that should receive special consideration included the thallide esters, in addition to thallic anhydryde, the C8 to C30 hydrocarbons, and, to some extent, the polynuclear aromatic hydrocarbons, even though there was very little analytical information on these from water quality studies or water analysis studies. Others included halogenated methanes, in addition to chloroform and carbon tetrachloride.

Largely through the efforts of Mr. Junk, the committee sorted out the compounds and classes of compounds which we considered, the approximate concentration that had been reported in USA drinking supplies, and a very rough estimate of the distribution in the different finished drinking water supplies. Carbon tetrachloride, for example, occurs at fairly low concentrations and is not widely distributed. Chloroform, on the other hand, has the possibility of a very high concentration and a wide distribution. For the most part, when we moved away from the halogenated compounds, there was relatively little information to allow serious considerations.

I would like to discuss and to summarize the conclusions of the study group. First, I will quote the summary of the study group and then go on and discuss the matter further. And, I quote, "An assessment of possible human health risk, associated with consumption of drinking water, contaminated with low concentrations of organic chemicals, depends on at least three factors: the adequacy of analytic methods for identifying and measuring the contaminants, and the scope of their application; the existence and adequacy of toxicological data on the contaminants; and the extent to which appropriate epidemiologic studies have been conducted to test a hypothesis of association, derived from the water quality data and the toxicological data." The study group addressed itself to these issues with carcinogenesis as the toxic effect of primary concern. It is recognized that a complete assessment of the possible risk should include those risks associated with exposure to other contaminants such as pesticides, asbestos, and inorganic chemicals, which were explicitly excluded from the charge to the study group.

It was the judgment of the study group that current methods of extraction, identification, and measurement, are available and adequate for field surveys of the identified drinking water contaminants of major toxicologic concern.

It is highly unlikely, however, that the majority of drinking water purveyors would have available sophisticated equipment and trained personnel sufficient to provide monitoring of the individual contaminants on a routine basis. It would be desirable if procedures could be developed to permit routine monitoring of groups of potentially harmful chemicals. Any trends in contamination of a water supply suggested by a chemical group monitoring program might then be subjected to more detailed study.

The study group also expressed concern that the chemicals which have been measured account for only a few percent of the total organic content of drinking water. Thus, attempts to evaluate the health risk of contamination may mistakenly be directed toward identified, potentially toxic compounds, while other compounds or groups of compounds, perhaps of equal or greater toxicologic significance, go undetected.

Furthermore, attention has been focused on the concentrations of contaminants in drinking water itself, while a complete analysis of the problem would also require analytic chemistry data on exposure to these chemicals by ingestion of food and beverages processed with contaminated water, and on other possible exposures resulting indirectly from environmental redistribution and possible biomagnification of the chemicals in food organisms which also can consume the water in question.

On the other hand, the study group felt that other unrelated sources, such as occupational exposures, to some of these chemicals in question would likely contribute a much greater potential intake than the consumption of drinking water. With respect to assessment of health risk exposure to the specific contaminants identified in the charge to the study group, it was concluded that some human risk exists.

This conclusion was reached on the basis of evidence that some of the compounds, particularly chloroform, are widespread contaminants of USA drinking water supplies, and that studies on laboratory animals indicate that chloroform produces hepatomas. It should be emphasized that experimental carcinogenesis data for chloroform are extremely limited, although support for its tumorgenicity is reinforced by more extensive studies demonstrating carcinogenic action of the related compound, carbon tetrachloride. These two compounds probably act by a similar mechanism, to produce hepatomas. Carbon tetrachloride, although occasionally identified as a contaminant of drinking water, occurs generally at much lower concentrations and is much less widespread as a contaminant than chloroform and related trihalogenated compounds.

Benzene has now been clearly established to be carcinogenic in experimental animals, although epidemiologic and clinical studies, largely of occupational exposures, suggest its possible carcinogenicity. Certain halo ethers, and polynuclear hydrocarbons, have been demonstrated to be carcinogenic in laboratory animals and have been identified in drinking water. To the very limited extent that they have been measured, the data available to the study group indicated that the potential human dosage of these compounds from ingestion of drinking water will generally be considerably less in absolute quantities than would be the case for chloroform.

However, the study group notes the existence of local situations in which this generalization would not apply. The study group felt that for all the compounds reviewed, the carcinogenicity data and experimental designs were generally either inappropriate or below the standard of current toxicologic practice and protocols for carcinogenicity testing. Additional well designed experimental studies to determine the carcinogenicity of lifetime exposure of ingestions to some compounds would be needed.

Data from epidemiologic studies on the contaminants of primary concern to the study group are very limited and the designs of the studies are generally inadequate for conclusive assessment of health risk. The recent studies alleging an association of high cancer incidence in New Orleans with consumption of contaminated drinking water are considered by the study group to be hypothesis formulating studies, but they should not be interpreted to have established a causal relationship. Numerous other variables might explain the apparent associations. Indeed, experimental toxicology studies suggest that if there were a carcinogenic risk, increased liver cancer would be a probable finding. This, however, was not revealed by the epidemiologic studies.

Recent studies of 79 cities of the National Organic Recognizance Survey have identified many water supplies in which some suspect halogenated organic compounds occur in higher concentrations than in New Orleans water or, in other words, in other cities. It should be possible, therefore, to test further the hypothesis formulated for New Orleans water and cancer in other cities that have a completely different set of variables from those of New Orleans.

In summary, based upon recent reasonably extensive water quality data from many USA water supplies and on extremely limited data from experimental carcinogenesis studies, the study group concludes that there may be some cancer risk associated with consumption of chloroform in drinking water.

The level of risk estimated from consideration of the worst case and for the expected cancer site for chloroform, the liver, might be extrapolated to account for up to 40 percent of the observed liver cancer incidence rates. A more reasonable assumption, based upon current water quality data which show much lower levels than the worst case in the majority of USA drinking water supplies would place the risk of liver cancer much lower, and possibly nill.

Further, it is emphasized that both the experimental carcinogenic data and the mathematical and biological extrapolation principles used to arrive at the upper estimate of risk are extremely tenuous. Epidemiologic studies do not, thus far, support the conclusion of an increased risk of liver cancer. Although, hypothesis formulating studies in southern Louisiana suggest the possibility of an association with contaminated water and overall high cancer incidence.

Critical definitive tests of this hypothesis have not been conducted, although some other organic contaminants contained in the charge to the study group have carcinogenic potential, the cancer risk to man is judged to be minor because of their low concentrations and/or infrequent occurrences in drinking water.

From our considerations, therefore, we concluded with an admittedly incomplete review of all the literature on chloroform and the other compounds, considering both the experimental carcinogenic data and the probability or possibility of exposure, that chloroform, perhaps, represents the greatest risk.

Chloroform, interestingly, occurred in finished drinking water in higher concentrations in several cases than occurs in the raw water sources from which the finished drinking water supplies were prepared. This indicates that chloroform, along with certain other low molecular weight halogenated compounds, may be formed in the chlorination process. There were wide variations in different cities, with the concentrations formed, even though chlorination was the disinfection process.

The summary I have just presented adequately indicates the reasons that we omitted or minimized the risk of certain of the other organic chemicals. I should point out that in spite of the quality of the data that we had to deal with, we did try and arrive at some estimate of risk. This was essentially the experimental carcinogenicity data on chloroform.

I believe there was one repititious study, by the same

authors, in 1944. A series of high doses were given intergastrically in thirty total doses every four days. It represented only about one-third of the lifetime of the mice and perhaps a little less than this. At the higher doses, there were other kinds of chemical toxicity which overrode any possible tumor effect and all the animals died at the 1.6 ml/k dosage. There was a sex difference in the particular strain of mouse used, for the males, at least, were much more susceptible to the acute neurotic action of the chemicals, and many died.

At the level of 8 ml/k, even though there were only four animals surviving at the time, they could be analyzed for tumors, all four of those surviving had tumors. At the .4 level, all of the surviving animals (in this case they were all females) had liver tumors. There was a dose at which no tumors occurred.

The one unique thing about this study was the fact that we had an effect level and a tested and reported level which produced no effects with respect to the tumor dose. We had no such data for most of the compounds. This is really not a very strong basis for tumorgenesis, but it was reported and we had no reason to question this. This was supported by the fact that carbon tetrachloride, in which more extensive studies (although not with any "no effect" dosage studies) had demonstrated in at least three species of mammals, tumor formation associated with exposure to high dosages. There were, however, no "no effect" dosage studies.

I might add, that recently the National Cancer Institute conducted studies with mice and rats which indicated that at the maximum tolerated dose and at one-half of the maximum tolerated dose, hepatomas were observed. I believe in both species they observed kidney tumors. In rats, it was about 12 over 13 percent. It was with this type of data that our biostatisticians attempted to estimate the risk for man.

But, this is case one, in which we considered the lowest dose at which all animals that survive develop tumors. Case two considers the lower doses, neither of which developed any tumors. Case three is an extrapolation using the most severe kind of mathematical approach one could select - the linear extrapolation method.

Additionally, certain assumptions were made. We used the principle that relating dosages from one species to another could best be done by consideration of surface area, so there was a conversion factor of about 14 from body weight to surface area of mouse to man.

We made assumptions that a dose could be expressed in terms of a milliliter per unit of body weight per day in order to give a daily dosage rate instead of the experimental rate that was actually used. This, then, was estimated to be equivalent to a certain dosage in a 70 kilogram man in terms of milligrams per kilo per day. This, then, expresses the water concentration that would be required in milligrams per liter to give a man dose equivalent to that which was related to the mouse dose, assuming that man drank four liters of water per day. Some would say that is too much, but others would say that is not enough. Nevertheless, that is what we used.

We could, therefore, get points estimating dosages that would cause a certain incidence in the first two cases, and in the third case, by extrapolation on a linear scale. Thus, what we have are confidence limits, estimated statistically for the case in which there was a 100 percent incidence in mice. This is the upper confidence estimated statistically from the zero incidence, and the upper confidence is estimated at 15 percent incidence. With the linear extrapolation down to zero, one can determine what would be the incidence at the worst contamination level that we were aware of then, 300 ppb.

With this drinking water by back calculating the incidence came out as an estimate of a tenth of a percent, which is, if one starts looking at it in terms of millions, a fairly large number. This could then be extrapolated down to any concentration desired. We recognize that this is taking the most extreme situation, and there is no reason to believe that there is no threshhold. There is no reason to believe there is not some kind of a curve or linear function to these dose response relationships.

We were looking at this from the standpoint of what might be the worst case, while recognizing we were working with inadequate data.

A summary of results, primarily of the studies by Harris and Associates of the Environmental Defense Fund, shows a statistical relationship of regression coefficients for the incidence of cancer for the variable, representing the proportion of drinking water from the Mississippi River. Certain coefficients are considered significantly different from zero at the 5 percent level. The regression coefficients have the dimensions of deaths per one hundred thousand per year, percent of water from the Mississippi River. That is finished water derived from the Mississippi and treated.

Therefore, a coefficient can be interpreted as an attributable risk in units of deaths per one hundred thousand per year on the basis of one hundred percent of water from the river, as compared to zero from the river in, presumably, ground water, which would not have the same finishing treatment.

The word "attributable" was used, as is common in epidemiologic literature, to refer to a difference in the rates between two exposure categories without implications that the difference is causally related to the exposure difference. The epidemiologists on our study group reviewed this and concluded that although this is a suggestion from which one might develop a hypothesis, it does not demonstrate a causal relationship. But, it is interesting to note that there was a non-significant, never negative, correlation for liver cancer. At the time we did the study, the report from the NCI was not available, but it is interesting to note that there was a positive correlation for genital urinary tract cancer in white males and non-white females.

In conclusion - I presented to you the summaries and conclusion of the study group; I have tried to illustrate the problems of evaluating health effects of contaminants in water, from inadequate data.

I think the study group felt that we were using chloroform, an estimation of risk here, as an illustrative example of what you might do to estimate risk. The study group does not conclude that chloroform is producing 40 percent of the hepatomas, but that is what the upper estimate of risk in the worst case might indicate.

Now - what has happened since that time? For one thing, we have the NCI report that I alluded to. Then again, and this is only anecdotal, I have heard that at least in one, and possibly two or three sites, concentrations of chloroform in treated, finished waters, have been measured at levels of eight and nin hundred parts per billion, instead of the 300 parts per billion. I have not seen the data, but I have heard people involved in the studies in the agencies say this.

Secondly, the list is constantly growing, whereas when we started, the first list we received contained 162 compounds - before we completed our review, we had 187 compounds. The June First list had 221 compounds, and the September First list had 235 compounds.

Certainly, the analytical capabilities are exceeding the rate of our toxicological and health evaluation capabilities.

What kinds of compounds are among these new ones that are being added to the list? I have selected a few of the some 65 additional compounds to show you the nature of some of these.

What is called bromomethane - I presume this is methyl bromide, but I must confess that for me, nomenclature gets a little questionable when I get a list with just one name on it. Nevertheless, I assume it is methyl bromide. Dr. Mrak will tell you that in California methyl bromide and chronic poisoning by methyl bromide, used as a grain fumigant, is a serious problem. Perhaps the most serious problem in California is related to pesticide use. Hence, I begin to wonder about methyl bromide and whether there should be concern about chronic neurological effects. I do not know what these levels would be. An acetylene bromide and an acetylene chloride were on this September 15 list, and not on the earlier list.

Acetylene is a very reactive compound, so we would be concerned about its presence. Vinyl chloride, which now I think everyone accepts, has a potential for carcinogenicity which also has been detected. Acetylene chloride is a closely related compound. Studies in our laboratory indicate it is much more toxic than either chloroform or carbon tetrachloride. I do not know that it is carcinogenic. Dr. Coulston is worried about cyanides in an organic matrix, and their release on hydrolysis in an organisms or in water, perhaps under certain pH concentrations.

With respect to isocyanic acids, we do not have room for complacency. On the other hand, I think, as has been alluded to previously, when we face the reality of acute infectious diseases as they are faced with in Latin America, the choices would probably be quite clear. We must take steps to protect against these.

When we are not faced with that kind of risk, then we must consider the problems that Mr. Junk raised a moment ago about alternatives. If you do not use chlorination processes, where do you go? To the chloramines which produce hemolysis? To ozonization processes, which may produce epoxides of even more concern, or hydrocarbons of equal or greater concern from a carcinogenic standpoint?

DISCUSSION

DR. COULSTON: What should we do? Jokingly, I think the most logical thing to do at this stage is to ban water. I don't want to take the risk of ingesting all these dangerous chemicals that are in water! Since safety testing of ingredients in water is a part of this discussion, the study that was described by Dr. Murphy, the mouse tumor study, doesn't necessarily mean cancer. We are talking about liver nodules, or hepatomas. A hepatoma simply means a benign growth.

Now, there are many who are saying that hepatomas should be considered, in animal tests of this sort, as an indication of carcinogenesis. The question of what is a chemical carcinogen is being debated hot and heavy right now. Some of the people in certain organizations of our government are pushing in that direction. Most pathologists agree on how a cancer diagnosis should be made. I hope regulatory agencies do not try to rewrite the textbooks. If they succeed, if the concept that a tumor should be considered as a cancer, if this ever became part of a regulatory system, obviously we would have to ban water (food) or take all these chemicals that produce mice hepatomas out of our supply. We can't take them all out. There is even the question whether some of these chemicals have ever actually been shown to exist in the water. But the fact is, that something is in the water. We have a right and a duty to delineate it, to try and find out what it is. What a task to have to ask Dr. Murphy and his committee, to debate an issue of this sort on the skimpy amount of evidence that was available to them. They are to be congratulated. I might add, that the incidence of liver cancer in the United States and Europe has been going down steadily for the past twenty years, not up.

It is not true, however, in other parts of the world where liver cancer is going up. I refer to Southeast Asia, primarily, and Africa.

DR. GOLBERG: I think the two issues I would like to raise in connection with what Dr. Murphy has said are: first of all, there has been a natural experiment going on for at least several generations, in the form of toothpaste containing considerable amounts of chloroform, which have been used by large populations in Europe, and to some extent, in this country.

It has been calculated that from the use of such toothpaste, partly through ingestion, but mostly through absorption from the mouth mucosa, milligrams of chloroform are absorbed everyday - many milligrams. And, it would be interesting if somebody could carry out in epidemiological survey to see whether this really extreme ingestion of chloroform, as well as its use for many years in cough syrups and a variety of pharmaceutical over-the-counter preparations, had any affect of the sort we are talking about.

The other point is there is an attempt to establish as a fundamental principle of carcinogenesis, that irrespective of the target organ in animals in which cancer appears as a result of exposure to a compound, cancer in man could appear in a totally different organ. Now, I am not saying that that is not true. There are compounds and occasions when this does happen. But, I would say that in the general, with these chlorinated compounds where the target organ is the liver and in man, also, we know that toxic effects are manifested in the kidney. I would be very surprised if, in fact, they were producing tumors in the brain or central nervous system, or the pancreas or something like that.

So, I think one has to be very careful about these generalizations which are being made and in which there are efforts to establish these generalizations as facts.

DR. SUESS: It is a very trivial question, but since I am not from the United States, and since I heard last week in Las Vegas quite a bit about studies performed in the water near New Orleans, and I heard Dr. Murphy talking about the study in New Orleans, I wonder whether there is a particular reason that all these studies have been performed in the drinking water of New Orleans? If I were some one from the press, after hearing these three days here and last week in Las Vegas, I would be very unhappy to hear what is wrong in the drinking water of Las Vegas, and I am sure it is full of bad chemicals. It is a very trivial question, but I have heard so much about New Orleans, that I wondered what is the special reason for a study there?

DR. MURPHY: As far as I can determine, the reason was that through the interests, perhaps of health officials in New Orleans, perhaps through public interest groups, there was a study made to look at the incidence of cancer. I will take that back - there have been city-wide and country-wide studies and reporting of cancer incidence throughout the country, and it has been known that New Orleans had an overall high cancer rate, for many years. And, New Orleans was among the first six study cities, I believe, that was included in the EPA's rather broad examination of chemicals.

And, I think it is just a case of a juxtaposition of interests, already available incidence data, and some new chemical analysis data. I stand to be corrected on that, because I am making certain assumptions. Certainly now, however, with these kinds of considerations, it seems to me that there could be identified certain communities that ought to be studied. And, I might add, that I grew up in that area, drinking water occasionally. Fortunately, I lived on a farm and drank most of my water out of the well. But the nearest town was the town that appeared with the highest concentration of any on this 80-city study, Huron, South Dakota.

Huron, South Dakota is a small populated area. It is a rural area. You probably would not have enough people to really do a good epidemiological study.

Indeed, there are many other sources of exposure to chloroform, and in deed, maybe cough drops and toothpaste may be responsible. An epidemiological study on those is the only way to evaluate the problem of chloroform, irrespective of the water situation. I am not an epidemiologist, but I gather that something with a low incidence, rather nonspecific, is difficult to detect and sort out, such as vinyl chloride, etc.

As far as questions of hepatomas and whether they are cancers, we sort of set our guidelines at the beginning, that we were not going to argue this question which cannot be agreed upon by oncologists and pathologists. We would accept this as an indication and we would consider it as risk of cancer.

DR. SUESS: Two points: Dr. Murphy read to us the summary statement of this specific study, which said, among other things, that a lot of the chemical substances studied were in a rather low concentration and, therefore, cannot be considered with respect to other sorts of contamination or pollution, and so on. I am not quoting, but I am just trying to repeat what I understood. Now, I have no reason to doubt the statement, and I, myself, would be very careful to make other types of statements like this.

But, I would think it is important to realize that there may be always people or groups that do not see the whole picture, but really, sometimes take words or sentences out of context. I, myself, was exposed to such situations, and therefore, I think it important to emphasize that this is not an invitation to continue to add unnecessary pollutants into the water system or air or what have you.

And I think the principle of trying to keep concentrations of pollutants, and other substances, as low as possible is most important. The natural concentration is an important thing to establish, which means we must indicate a maximum permissible concentration. That is true for radiation and radiochemicals, and it should be true here. This principle can be used by certain industries, civil groups, etc., in either direction.

A second point is the question of threshhold. I am not

considering myself an expert in the field, but from what I understand and have read, as written by experts, as far as carcinogenicity is concerned, there may be no threshhold at all at one point.

DR. COULSTON: There are more and more experts in the world who are accepting threshhold levels for so-called chemical carcinogens, especially in mice. And they are more and more agreeing that there is a dose response to a so-called chemical carcinogen. So, I don't know where you get that statement. Your example of radiochemicals and radiation assumes a no-effect and threshhold level, or otherwise, we could not permit their use or would have to remove them from the environment.

DR. SUESS: A combination of many doses on a long-term basis are more significant than a single dose of accumulative amount which means, here again, we have something which we cannot really check on a long-term basis as easily.

DR. MURPHY: We had concern about some of the things that we said in our report being taken out of context. Quite honestly, I think it was handled fairly well. I don't believe anyone blew things out of proportion from what we said, and it could have been done out of context. Certainly, we indicated that we felt that because certain chemicals occurred at lower concentrations or lesser frequency, they didn't represent as much a risk. I would not want this to be misunderstood to mean that something of a low concentration could not represent a very serious problem. But, relative to the experimental data and the dosages required to produce certain effects, chloroform stood out. In our full report there is a comparison. I indicated we had positive data for several compounds, such as carbon tetrachloride, chloroform, and some of the halo ethers. We looked at this and we had positive data. We did not have negative data, and the biostatistician said we can't really deal with this in the same way.

But, we could do a lesser thing. We could calculate the ratio between the dose that produced 50 percent tumors in mice or rats, to the possible dose that man might get, drinking four liters of water per day, at the highest concentration that has been detected. In every case that we considered, chloroform came out with the smallest ratio, some of them did come out with 100,000 X safety ratios. I think maybe the chloroether had that kind of a ratio, but this is explained in some detail in the report of the study group.

DR. EGAN: I cannot comment on the whole of the 267 or 235 chemicals and I only hope that Dr. Murphy and his _ad hoc_ study group would look critically at the evidence for the establishment of the identity of these chemicals. The conclusion depends a bit on the level of the chemical found. But, basically, they are subject to the same kind of philosophy of confirmation of positive identity, as the things we have been talking about earlier. My first reaction would be to look very critically at the evidence for the occurrence of those chemicals, at least some of them, because others, I think, might already have been established, and so one would have already looked criticially. What I am very slightly worried about is the fact that the earlier remarks by Dr. Hutton, for example, and others, mentioned that research is unbalanced and it is too sophisticated and the methods are too sensitive - fair enough.

One of the conclusions of our Forum, was the fact that any firm view depended on the adequacy of analytical methods, as far as toxicological data. You must tell the analytical chemist what you want him to do, otherwise, he just goes on trying harder and harder until you stop him.

DR. JUNK: I think that this is the point. I, myself, read gas chromatograms, although I am not a chemist. But the chemists bring them in to me and say, "Oh, look at this littly tiny blip. I think that's dieldrin. I think that's something." And, I answer, "What do you mean, 'you think?' Do you know or don't you know?" Well, "Let's do a mass spec." I would say eight out of ten times the answer comes back and it is not what it was thought to be - it's something else.

Now my whole history is in mass spectrometry, and I have a prejudiced viewpoint regarding its ability to identify organic chemicals, but, I think it is a safe statement that no other single analytical technique approaches, in power, the ability of mass spectrometry to identify small amounts of organic compounds, with the right kind of mass spectrometry.

DR. COULSTON: This is, of course, perfect, and it means that it is far better for someone to be certain about chemicals appearing in water before they proclaim they have. They must go to their peers, re-evaluate, and make certain, as you have said in your case, you have done. All of us would feel better then, to say that these compounds do, indeed, exist. And we should do something about them, either decide they are no risk or say they are a risk and take them out or restrict or ban them from water. That's all there is to it.

DR. KORTE: You know, I am an optimist, and I would like to state, always, everything can be measure if one has the right methods, the right methodology. If one has enought time, it might take a lifespan to identify one compound, for instance, but this is very seldom possible. Therefore, one always has to compromise, and in compromising, we haven't got enough time to get an answer. But principly, everything can be done in chemistry.

DR. GOLBERG: I would like to express my curiosity about the other 95 percent of the organic material that is present in water, which has not been accounted for. It reminds one of the story of the man who was looking under the streetlamp for his wallet. Even though he had not lost it there, but that was where the light was. If a compound has halogen in the molecule, that is why we are able to pick it up and identify it. But, if there are the kinds of things from natural sources which could conceivably constitute a carcinogenic hazard after chlorination, after peroxidation, or something like that, we are totally unaware of these. And so, I think the larger part of the problem has not even been touched upon. And that may prove to be the most important work.

DR. COULSTON: This is absolutely true. We touched on this with lead and other atmospheric chemicals where we said we cannot identify the organic phase of the metal attachment.

You heard from various speakers that there are a great number of other substances present in water. We are only looking at a small percentage of what must be there. Why? Because the methods are available, as has been said, to measure these chemicals. I would be more worried about the 95 percent of the chemicals in the water that we don't even know how to analyze for, rather than the one or five percent that are known. Unfortunately, the detectors have not as yet been designed to detect all of the chemicals in other groupings, in other molecular configurations, and with mass spectrometry, we are limited to chemicals below 5000 molecular weight. The average machine, below 1000.

DR. JUNK: You are provoking the chemist to find more, because there was a time, 10, 20, 30 years ago, when he hadn't discovered many of the things you have been talking about these last three days.

DR. COULSTON: Exactly the point, and you say it well. We must know and we must not be afraid of the unknown.

DR. JUNK: By way of analogy, I think the situation with water is somewhat similar to drawing an analogy to the seashore, which contains many pebbles of sand. I will refer to these being the small molecules. And, many boulders, now in chemical characterization, particularly of biological materials and many natural materials, we have not begun to attack the boulders, yet. We are doing a fair job on the pebbles. I think that can be extrapolated to the water situation, if the materials are, and I will put several qualifications, generally non-polar and of low molecular weight; I think we do a very good job.

If I were pressed into making an estimate, I would say that the 90 percent unidentified components represent in number of compounds, less than what we have identified up to this point. But, if we are looking for characterizations of, let's say the soil, chemists have been trying for twenty years to characterize the first humic acid. We already know that there are large amounts of humic acid, and it represents a good share on a weight basis, of this 90 percent unidentified components. We do know at this stage, our natural humic materials. And we must also include polysaccharides and a variety of lignant type materials, too.

DR. COULSTON: You see the real problem, now, even with mass spec, there are limitations. What is the maximum molecular weight you can go to?

DR. JUNK: You can measure chemicals around 1500 molecular weight, above that, you're in real trouble.

DR. COULSTON: The difficulty is we can only identify certain things that are relatively small in molecular weight, and when it comes to large molecules, which are probably more dangerous to public health, we cannot do it. As has been said, we are just scratching the surface of an iceberg. We can only do so much, and we need more and more research. We need more sophisticated equipment, we need time before we know whether a chemical is present, and if it is harmful.

I am never worried that the chemist is going to find something down to the level of one molecule, or even a side-chain of a molecule. This never bothers me. If we know what is present, concepts of toxicology will help us decide whether or not it is safe.

The point is that we have to use, at the present time, reasonableness, benefit-risk relationships, socio-economic considerations, to decide and we need and must have the facts. Without these considerations, we are going to be in one terrible situation, for as Dr. Junk said, we tend to ban those things we know the most about.

DR. EGAN: The only point I wanted to add was that to bear in mind, first of all, as Professor Korte told us right at the beginning of the meetings, that water is a natural substance. There are many kinds of natural waters. Many natural compounds are high molecular weight and I would not like it to be thought that it follows automatically that any unidentified high molecularic weight materials in water are not natural substances.

DR. COULSTON: Very good point. I think we have come a long way in the last three days. I don't think we are more confused than when we started. I do believe we have all learned a great deal, some more than others.

The speakers and participants have given freely of their thoughts and experiences. It has been a great pleasure for me to have all of you here from different parts of the world, to sit around the table and contribute so much to this outstanding conference. Remember, only through a full exchange of ideas will progress be made.

My thanks to all the participants.

Index

A

B

C

E

F

H

I

L

M

N

O

P

R

S

T

U V

W

A
B 7
C 8
D 9
E 0
F 1
G 2
H 3
I 4
J 5

305337342U